Leckie×Leckie
Scotland's leading educational publishers

ESSENTIAL EXAM SKILLS

Higher
CHEMISTRY
GRADE BOOSTER

Tom Speirs

© 2017 Leckie & Leckie Ltd

001/22062017

10 9 8 7 6 5 4 3 2 1

ISBN 9780007590841

Published by
Leckie & Leckie Ltd
An imprint of HarperCollins*Publishers*
Westerhill Road, Bishopbriggs, Glasgow, G64 2QT
T: 0844 576 8126   F: 0844 576 8131
leckieandleckie@harpercollins.co.uk
www.leckieandleckie.co.uk

Commissioning editor: Clare Souza
Project manager: Rachel Allegro

Special thanks to
Louise Robb (copy edit)
Dylan Hamilton (proofread)
Jouve India (layout and illustration)
Paul Oates (cover)

Printed in China

A CIP Catalogue record for this book is available from the British Library.

Acknowledgements
All images © Shutterstock.com

SQA questions reproduced with permission (solutions do not emanate from SQA), Copyright © Scottish Qualifications Authority.

Whilst every effort has been made to trace the copyright holders, in cases where this has been unsuccessful, or if any have inadvertently been overlooked, the Publishers would gladly receive any information enabling them to rectify any error or omission at the first opportunity.

Contents

Introduction

This chapter covers:

- What this book is for
- How to use this book
- When to use this book
- Course design and assessment
- The exam
- What you should already know
- Areas of difficulty
- The Chemistry Data Booklet
- Chemical reagents

What this book is for

Chemistry has traditionally been viewed as one of the more challenging courses, with new concepts to understand, and both practical and problem–solving skills to develop. However, as well as being challenging the course can ultimately be tremendously satisfying, giving those who study it a better understanding of their environment, and allowing students to develop informed opinions on the many issues that face our society.

This book is designed to support the work that you will have covered in class or at college and to help you gain a better understanding of the material covered in your course. Better understanding should allow you to achieve your full potential in the course assessments.

Success always comes at a price. Whatever we choose to do, be it academic study or playing sports or participating in performing arts, there is a key to success. It is simply this – application, application, application!

A spectator once suggested the golfer Gary Player had been lucky by holing three bunker shots in a row when practising for a tournament. Player simply replied, 'The more I practise, the luckier I get'.

Just as it is true for sports so it is true for our studies: the more we practise, the better we become. The more time we spend on an activity the more confident we will be in our own abilities. The more time spent reading course notes and looking at the questions that appear in the Higher papers, the better you will likely be when it comes to the course assessment.

This book doesn't revisit all the theory that you have covered in the course. You should have course notes that adequately do this or there are various good textbooks that cover the course theory. We will, however, highlight crucial points as we tackle the various questions. We will focus on the different **types of questions** that you will be asked and the areas of the course that candidates find most demanding. If you achieve success in these areas then hopefully you will give yourself an improved chance of achieving higher marks and grades in the course assessments.

How to use this book

Understanding the areas where we have secure knowledge and good understanding, and those where we have weaknesses, is the key to making progress. We all need feedback on how best to improve. The first port of call must always be your teacher or lecturer who can clarify those points you didn't quite understand. Asking your teacher or lecturer questions will give them a real insight into your specific needs and help them to target support to you as an individual. This book will support the dialogue you should be having with your teacher. It will help you understand those areas where you can be confident in your understanding and abilities and help you identify those areas where you may have weaknesses and where you would benefit from asking questions that will help clarify matters.

It is not necessary that you should read the book starting at the beginning and working your way through in a sequential manner. You will benefit from reading every chapter but, as you become more aware of those areas where you would like to improve, you will be able to identify those chapters that are most relevant to you. The various chapters will help you develop strategies that will allow you to tackle every type of question that might be given in the final exam or create a suitable report on your assignment.

When to use this book

The book is intended to be used throughout your course. Very soon after you start your course you are likely to face questions, either in class or as homework, that will be similar to those you will face in assessments of the various Key Areas, prelim exams or in your final Scottish Qualifications Authority (SQA) assessment. Becoming familiar with the type of questions you will face, early in your course, will help you achieve your full potential in these Key Areas. It will be important as you approach prelims or the final exam that you spend more time systematically preparing for these exams. Using this book will help ensure that you cover all the areas that you need to, in order to be prepared for these exams.

Course design and assessment

Full details concerning the course content and the assessment specifications for both the final exam question paper and the assignment can be found on the SQA website, www.sqa.org.uk. In the search box at the top of the home page, simply type in 'Higher Chemistry'. You can then find the section of the website that gives you access to materials relating to Higher Chemistry. These include course support notes, specimen question papers with mark schemes, general coursework information giving details of how your assignment will be marked, and the Chemistry Data Booklet. Although much of the information is written for teachers, you are allowed to download the information for your own use as a student. The information given below is taken from the SQA website.

The main aims of the Higher Chemistry course are for learners to:

- develop and apply knowledge and understanding of chemistry
- develop an understanding of chemistry's role in scientific issues and relevant applications of chemistry, including the impact these could make in society and the environment
- develop scientific inquiry and investigative skills
- develop scientific analytical thinking skills, including scientific evaluation, in a chemistry context
- develop the use of technology, equipment and materials, safely, in practical scientific activities, including using risk assessments
- develop planning skills
- develop problem-solving skills in a chemistry context
- use and understand scientific literacy to communicate ideas and issues and to make scientifically informed choices
- develop the knowledge and skills for more advanced learning in chemistry
- develop skills of independent working.

Throughout the course it is intended that you should develop an understanding of the impact of chemistry on everyday life, and the knowledge and skills that will allow you to reflect critically on scientific and media reports and to make your own judgements on the many issues that affect our society.

The course assignment helps develop investigative skills and communication skills. The course also helps develop literacy and numeracy skills.

The course allows you to study chemistry in contexts that are both relevant and up to date. The course is organised under four headings:

- Chemical Changes and Structure
- Nature's Chemistry
- Chemistry in Society
- Researching Chemistry

The exam

The national examination is divided into two sections:

- Section 1 contains multiple choice questions
- Section 2 contains structured questions requiring written answers.

Over the paper the mark distribution between Knowledge and Understanding and Skills is approximately:

- Knowledge and Understanding: 70–75% of the marks
- Skills: 25–30% of the marks

Knowledge and understanding

The Higher Chemistry course requires learners to retain and understand chemical information and to integrate that knowledge to give understanding of other areas of chemistry.

Throughout the course you will find that knowledge gained in one topic is needed to fully understand information in other topics. For example, hydrogen bonding covered in Chemical Changes and Structure is needed to understand why alcohols have high boiling points in Nature's Chemistry.

An important aspect of chemistry is the requirement to assimilate knowledge and apply this knowledge in new situations. The ability to do this is reflected in many of the questions used in the final exam. For example, an exam question on atom economy or percentage yield covered in Chemistry in Society may be set in a context related to Nature's Chemistry.

Knowledge questions are subdivided into **3 categories**:

- questions that ask you to **recall information**
- questions that ask you to **apply knowledge** in new situations (some of these questions were previously categorised as problem-solving)
- questions that ask you to **give explanations** based on knowledge and understanding of the chemistry of the course.

Only about 12 of the knowledge marks are for straight recall of information. The majority of the marks are for showing understanding by applying knowledge and giving explanations.

Knowledge questions **generally begin with a command word**, such as *state; name; suggest; describe; calculate; explain*, etc.

It is important when answering questions that you look carefully at the mark allocation, particularly for questions that require explanations.

Some questions will simply ask you to explain and may only be worth 1 mark, other questions may begin with '**Explain fully**'. These questions are generally worth 2 or 3 marks. In these questions it is common for some guidance to be given to help you construct your answer.

Example
In the **2014 Higher (Revised) Chemistry paper Q3(b)** candidates were asked to '**Explain fully**' why the boiling point of hydrogen fluoride was much higher than the boiling point of fluorine.

The question was worth 3 marks. The question writer gave candidates a hint as to what was required in the answer by adding 'In your answer you should mention the intermolecular forces involved and how they arise'.

The area where most candidates failed to gain marks in the example given above was in not answering the 'and how they arise' aspect of the question.

Remember these three points when answering questions.

1. Read the entire question, and read it carefully.
2. Remember words written in bold are there to help you focus on what is required in your answer.
3. Look at the mark allocation for the question.

As we go through the examples in the book it will be important that you bear these three points in mind.

Skills

As well as gaining knowledge and understanding, learners are expected to develop data-handling skills during the course. These skills are also assessed in the end-of-course exam.

The data-handling skills that you will be required to demonstrate are your ability to:

- select information and present information appropriately in a variety of forms
- process information, using calculations and giving units where appropriate
- make predictions and generalisations from evidence / information
- draw valid conclusions and give explanations supported by evidence / justification.

Particular question types

There are particular question types that you should be aware of. These constitute a significant proportion of the marks available in the Higher Chemistry exam and therefore need to be practised thoroughly if you are to achieve your full potential in the exam. Separate sections of this Grade Booster book will deal with answering these types of questions.

It is all too easy to think *'I can't do ...'* and leave a question out but the key to success is to be more positive. Instead of *'I can't do...'* , think *'I'm having difficulty with..., I need to find out or I need to practise...'*.

There is no substitute for hard work – as has been said:

Hard work will bring rewards and you won't gain the rewards without the willingness to push yourself.

The weighting of particular question types

The Principal Assessor who constructs your paper is given particular criteria that have to be satisfied when setting the paper. This involves allocating a set number of marks to particular types of question. This means you shouldn't be surprised by the marks allocated to such areas as calculations, open questions and questions testing Researching Chemistry skills.

Calculations

There are two different types of calculation that appear in the Chemistry Higher exam: those that assess knowledge of chemistry calculations that appear in the course, such as Hess's law calculations, and calculations that assess general numeracy skills, such as the use of percentages or scaling factors in a chemical context.

The approximate mark allocation in the end–of–course exam is:

- Chemistry calculations   15 marks
- Numeracy calculations   8 marks

Open questions

The end-of-course exam contains two 'open' questions, each worth 3 marks. These questions are signposted, starting with the words **'Using your knowledge of chemistry...'**, and allow candidates to examine or consider a chemical situation and describe underlying chemistry relating to the situation. The 'open-ended' nature of the questions is such that there is no unique correct answer. Rather the candidates are awarded marks according to the understanding of the underlying chemistry:

- Good understanding   3 marks
- Reasonable understanding   2 marks
- Limited understanding   1 mark

Researching Chemistry questions

Approximately 10 marks in the question paper relate to assessment of practical skills associated with equipment and techniques described in the Researching Chemistry section of the course.

What should I already know?

Here's what the SQA says:

'Learners undertaking the Higher Chemistry course would normally be expected to have attained the skills, knowledge and understanding covered and developed in the National 5 Chemistry Course.'

Although you may not be examined on content defined at National 5 level, there are particular areas where your understanding needs to be secure since these will underpin your learning at Higher. These areas are:

- atomic structure, nuclide notation and isotopes
- chemical formulae
- balanced equations
- calculations relating mass, volume of solutions, concentration and moles
- reactions of acids
- volumetric calculations
- homologous series
- drawing and naming of branched–chain alkanes and alkenes
- enthalpy calculations ($E_h = cm\Delta T$)
- oxidation, reduction and redox.

If you are in any way unsure in any of the above areas, it will be worthwhile reading through National 5 materials and working through National 5 questions in the relevant areas. You can also ask your teacher or lecturer for materials that will improve your understanding in these areas.

Areas of difficulty

Each year the SQA publishes the Principal Assessor's report for Chemistry. This includes a section detailing areas where candidates found difficulty in answering the questions in the end–of–course exam. There are recurring areas mentioned in these sections. Questions that feature prominently are questions that relate to:

- bonding and structure
- intermolecular forces
- concept of excess in calculations
- combining and balancing ion–electron equations
- oxidising and reducing agents
- describing experimental procedures, particularly those relating to Researching Chemistry techniques
- drawing diagrams.

As these are areas where candidate performance tends to be weak, it therefore makes sense to prioritise these when preparing for your exam. There are many questions in past paper booklets that focus on these areas. Make sure you allow time to work through them.

Top tip

When you are preparing for the exam, it sometimes helps to adopt a themed approach, focusing on a particular area of your course work such as bonding and intermolecular forces or redox equations. Then, instead of working through a single past paper, find as many questions from different past papers and other sources that relate to the area and work through these.

The Chemistry Data Booklet

The Chemistry Data Booklet is a valuable source of information. The following pages are the ones most relevant to Higher.

Page 4 – *Relationships for Higher and Advanced Higher Chemistry*

This gives useful formula relationships particularly for % yield and atom economy.

Page 5 – *Names, Symbols, Relative Atomic Masses and Densities*

> Relative atomic masses are given correct to one decimal place for use in calculations where Gram Formula Mass of a substance needs to be used.

Pages 6, 7 and 8 – *Melting and Boiling Points; Covalent Radii; Electronic Arrangements of Selected Elements*

> Useful periodic tables when considering periodic trends for the elements.

Page 9 – *Melting and Boiling Points of Selected Organic Compounds*

> This table is useful when considering properties of members of a homologous series or when considering impact of functional groups on physical properties.

Page 10 – *Enthalpies of Combustion; Bond Enthalpies of Elements and Mean Bond Enthalpies*

> You are required to know the definition for Enthalpy of Combustion. This is energy released when 1 mole of a substance burns completely. You should also know that when using bond enthalpy values, bond breaking requires energy and is therefore endothermic (positive enthalpy value) and conversely bond making releases energy, i.e. is exothermic (negative enthalpy value).

Page 11 – *Ionisation Energies and Electronegativities of Selected Elements*

> These properties of the elements can be used to illustrate periodicity, i.e. the trends in the properties across a period and down a group.

> You are required to know that the first ionisation energy is the energy required to remove 1 mole of electrons from 1 mole of gaseous atoms. Very helpfully, an equation is given above the table if you happen to forget this.

Page 12 – *Electrochemical Series: Standard Reduction Potentials*

> This gives us the ion-electron equations for reduction reactions reading from left to right. To obtain the equation for oxidation of a species look for the species on the right and reverse the equation.

> For example the reduction of iron(III) is given by the equation

$$Fe^{3+} (aq) + e^- \rightarrow Fe^{2+} (aq)$$

> If the equation is reversed, the equation for the oxidation of iron(II) is obtained

$$Fe^{2+} (aq) \rightarrow Fe^{3+} (aq) + e^-$$

> As well as giving you the ion-electron equations, the table allows you to compare substances as oxidising agents or reducing agents. The strongest oxidising agents (electron acceptors) are found at the bottom left of the table. The strongest reducing agents (electron donors) are found at the top right of the table.

You need to spend time familiarising yourself with the contents of the Data Booklet. Don't just go to it when you need to look something up.

When you are doing periodicity in class, look at the different periodic tables or the table of ionisation energies on page 11 of the booklet, note the periodic properties, and try to think through why these occur. Think about why elements in groups might have properties in common and why properties such as ionisation energies vary across a period.

When you are looking at the effect of functional group in alcohols, aldehydes and ketones go to the information on page 9 of the booklet, look at the boiling points and consider why they are different. Just knowing what is in the Data Booklet can prove invaluable when under pressure in an exam.

Chemical reagents

A number of different chemical reagents are mentioned in the Chemistry support notes. You need to know what they are used for and the colour change associated with each reagent.

Reagent	Use	Colour change
Acidified potassium dichromate	Oxidation of primary and secondary alcohols, and aldehydes	Orange to green
Hot copper(II) oxide	Oxidation of primary and secondary alcohols	Black to red
Tollens' reagent	Oxidation of aldehydes; in order to distinguish aldehydes from ketones	Formation of a silver mirror
Fehling's solution	Oxidation of aldehydes; in order to distinguish aldehydes from ketones	Blue to brick red
Potassium permanganate	As self–indicator in some redox titrations	End point indicated by appearance of first permanent pink tinge when potassium permanganate is added

In addition candidates would be expected to know that starch gives a blue–black colour with iodine solution and that bromine solution is decolourised (red to colourless) by unsaturated carbon compounds.

Do you know your definitions?

> **This chapter covers:**
>
> - Chemical Changes and Structure definitions
> - Nature's Chemistry definitions
> - Chemistry in Society definitions

Knowing your definitions

Knowing definitions is often the key to being able to answer many of the questions in Chemistry. You may be asked to give a definition, such as: 'State what is meant by a free radical'. It is more likely that your knowledge of chemistry definitions will be examined in an indirect way when you are asked to apply knowledge of a definition to a particular situation. For example, when given an equation for a reaction you may be asked to state the type of reaction that is taking place.

We will look at, and in some cases explain, the definitions you need to know for Higher Chemistry.

Chemical Changes and Structure definitions

Activation energy
The minimum kinetic energy required by particles before collisions will result in a reaction taking place.

Activated complex
An unstable arrangement of atoms formed at the maximum of the potential energy barrier, during a reaction. The activated complex forms when particles collide with sufficient energy for a reaction to take place.

Electronegativity
The attraction of an atom for the shared electrons in a bond.

Fluorine is the most electronegative element. This is because the nucleus has a very strong pull on bonded electrons due to its small atom size. Atom size decreases across a period of the periodic table. Atom size and shielding by inner electron shell increases down a group. Electronegativities increase across a period and decrease down a group in the periodic table.

First ionisation energy

This is the energy required to remove 1 mole of electrons from a mole of atoms in the gas state.

The equation representing the first ionisation energy for an element E is:

$$E(g) \rightarrow E^+(g) + e^-$$

The second ionisation energy for an element would represent the energy to remove a second mole of electrons. The equation for this would be:

$$E^+(g) \rightarrow E^{2+}(g) + e^-$$

Van der Waals' forces

A general term used to describe the different types of force that can exist between atoms and molecules.

London dispersion forces

Weak forces that exist between atoms of noble gas elements and all molecules, due to temporary dipoles caused by the movement of electrons.

Permanent dipole – permanent dipole interactions

Intermolecular forces due to attractions between molecules having dipoles.

Hydrogen bonds

An extreme form of permanent dipole–permanent dipole interaction occurring when a hydrogen attached to a nitrogen, oxygen or fluorine atom is attracted to a nitrogen, oxygen or fluorine atom on a neighbouring molecule.

Bonding continuum

The gradual change in bonding type that occurs in going from pure covalent to ionic bonding.

Nature's Chemistry definitions

In Nature's Chemistry a number of different chemical reactions are defined. You must be careful since a number of them involve water.

Condensation reaction

A reaction in which molecules join by eliminating a small molecule such as water from between them.

ethanoic acid   methanol   methyl ethanoate   water

The formation of the ester methyl ethanoate from methanol and ethanoic acid is an example of a condensation reaction. The reaction can also be described as esterification, the formation of an ester. The arrows $\rightleftharpoons$ in the equation indicate that the reaction is reversible. Methyl ethanoate can break down to methanol and ethanoic acid. This reaction is described as hydrolysis.

Hydrolysis
A reaction in which molecules are broken apart by the addition of water.

Proteins breaking up into amino acids is an example of hydrolysis.

To identify the products of hydrolysis, first identify the bond that will be broken. In a dipeptide two amino acid units are joined by a peptide link.

1. Identify the peptide link in the structure.

2. Break the bond between the carbon and the nitrogen and add OH to the CO and H to the NH.

Be careful not to confuse hydrolysis with hydration, another type of reaction that you may have heard about. In hydration, water is added to a molecule but the molecule does not break apart.

Hydration (and dehydration)
Hydration is the addition of water to a molecule. The water adds across a double carbon to carbon bond. This type of reaction can be used to make ethanol from ethene.

Hydration

In dehydration reactions water is removed leaving a double carbon to carbon band. If propan–2–ol is dehydrated then propene is produced.

Dehydration

Hydrogenation (and dehydrogenation)
Hydrogenation is the addition of hydrogen to a molecule. Hydrogenation of alkenes produces alkane molecules.

For hydration and hydrogenation reactions, the ending '–ation' is used to indicate 'the addition of'. In the same way chlorination would be the addition of chlorine and bromination would be the addition of bromine.

Denaturing
This is a process in which proteins change shape when intermolecular bonds break, usually due to high temperatures or changes in pH. Enzymes are proteins and will lose their biological activity when denatured.

Oxidation and reduction
Oxidation and reduction are defined in a number of ways at Higher. In National 5 you would likely use a mnemonic such as **OILRIG** to help you remember the definitions:

OIL – **O**xidation **I**s **L**oss of electrons
RIG – **R**eduction **I**s **G**ain of electrons

The definitions you learn at Higher are consistent with this definition.

In carbon chemistry a simple way to recognise that oxidation and reduction are taking place is to look at the hydrogen-to-oxygen ratio of molecules.

Take a primary alcohol changing to an aldehyde.

ethanol   ethanal

Ethanal has a greater oxygen to hydrogen ratio (1/4) than ethanol (1/6). The oxygen to hydrogen ratio is increasing, therefore the ethanol is being oxidised.

This is consistent with our original definition as if we use our ion–electron equation balancing rules we end up with:

$$C_2H_5OH \rightarrow CH_3CHO + 2H^+ + 2e^-$$
ethanol   ethanal

We can see that the process involves loss of electrons and therefore is consistent with our original definition.

Reduction is the opposite of oxidation. In reduction there is a decrease in the oxygen-to-hydrogen ratio.

Free radical
This is an atom, molecule or ion that has an unpaired electron and is therefore highly reactive.

Free radical chain reaction
A reaction that can be defined in three stages: *initiation, propagation* and *termination.*

- *Initiation* – the stage where free radicals are formed when a covalent bond breaks, creating two species with unpaired electrons; for example:
$$Cl_2(g) \rightarrow 2Cl\cdot(g)$$
- *Propagation* – a stage in the chain reaction where a free radical reacts with another species, creating another free radical.

- *Termination* – the stage in the reaction where free radicals combine to form a stable substance.

Chemistry in Society definitions

Atom economy
This is a measure of the efficiency of a chemical process. Atom economy is defined as the percentage of reactants by mass that can be changed into the desired product during a chemical reaction.

$$\% \text{ atom economy} = \frac{\text{Mass of desired product(s)}}{\text{Total mass of reactants}} \times 100$$

Percentage yield
This is another measure of the efficiency of a chemical reaction. It is the percentage of the theoretical yield that is produced during a reaction.

$$\% \text{ yield} = \frac{\text{Actual yield}}{\text{Theoretical yield}} \times 100$$

Limiting reactant
When reactants are mixed but are not in the exact proportions indicated by the balanced chemical equation for the reaction, this is the reactant that determines the mass of product that can be produced. The other reactant(s) would be in excess.

Closed system
A system where mass cannot be added to or taken from the system. The contents of a sealed flask would represent a closed system.

Dynamic equilibrium
The point where the rate of the forward reaction and the rate of the back reaction are equal for a reversible reaction.

Enthalpy of combustion
The energy released when 1 mole of a substance is burned completely in oxygen.
The following equation represents the combustion of propane.

$$C_3H_8(g) \ + \ 5O_2(g) \ \rightarrow \ 3CO_2(g) \ + \ 4H_2O(\ell) \quad \boxed{\Delta H = -2219 \text{ kJ mol}^{-1}}$$

This means that when 1 mole of propane burns completely, 2219 kJ of energy will be released.

Hess's Law
This law states that the energy change for a chemical reaction is independent of the pathway between the initial and final states.

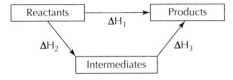

By Hess's Law $\Delta H_1 = \Delta H_2 + \Delta H_3$

Molar bond enthalpy
For a diatomic molecule, XY, the molar bond enthalpy is the energy required to break one mole of XY bonds.

Mean molar bond enthalpy
This is the average energy required to break a bond that occurs in different molecular environments. For example the energy required to break an oxygen hydrogen bond in an alcohol would be slightly different from the energy required to break the same bond in a carboxylic acid.

Oxidising and reducing agents
Oxidation is loss of electrons. An oxidising agent, something that will help oxidation to take place, is therefore a substance that will accept electrons. Because it accepts electrons, an oxidising agent is reduced during a chemical reaction.

Top tip

Always work from OILRIG.

In the same way, reduction is gain of electrons. A reducing agent, something that will help reduction to take place, is therefore a substance that can donate electrons. Because it loses electrons, a reducing agent is itself oxidised in a chemical reaction.

OXIDATION	Loss of electrons	OXIDISING AGENT	Electron acceptor
REDUCTION	Gain of electrons	REDUCING AGENT	Electron donor

Remember, the electrochemical series on page 12 of the Data Booklet allows you to identify the strongest oxidising agents and reducing agents. The strongest oxidising agents (electron acceptors) are found at the bottom of the table on the left. The strongest reducing agents (electron donors) are found at the top of the table on the right.

Volumetric analysis
The use of a solution of accurately known concentration to determine the concentration of another substance.

Standard solution
A solution of accurately known concentration.

Examples

Questions requiring you to state a definition tend to be worth 1 mark. Questions where you need to apply a definition may be worth more marks.

Q Which of the following equations represents the first ionisation energy of chlorine?

A $Cl(g) + e^- \rightarrow Cl^-(g)$

B $Cl^+(g) + e^- \rightarrow Cl(g)$

C $Cl(g) \rightarrow Cl^+(g) + e^-$

D $Cl^-(g) \rightarrow Cl(g) + e^-$

A If you know that the definition is the energy to remove an electron from an atom in the gaseous phase you will get the answer easily. Two options have an atom on the reactant side. In A the atom is gaining an electron whilst in C an electron is being removed. The answer is clearly C.

Q Which of the following equations represents the enthalpy of combustion of propane?

A $C_3H_8(g) + 5O_2(g) \rightarrow 3CO_2(g) + 4H_2O(\ell)$

B $C_3H_8(g) + \frac{7}{2}O_2(g) \rightarrow 3CO(g) + 4H_2O(\ell)$

C $C_3H_8(g) + 3O_2(g) \rightarrow 3CO_2(g) + 4H_2(g)$

D $C_3H_8(g) + \frac{3}{2}O_2(g) \rightarrow 3CO(g) + 4H_2(g)$

A Enthalpy of combustion is the energy given out when 1 mole of a substance is burned completely. The word 'completely' is important. All the equations show 1 mole of propane on the reactant side. Options C and D are quickly dismissed – substances don't burn to produce hydrogen! We're left with A and B. This is where 'completely' comes in: B is incomplete combustion – carbon monoxide is being produced, so the correct answer is A.

Q An oxidising agent

   A gains electrons and is oxidised

   B loses electrons and is oxidised

   C gains electrons and is reduced

   D loses electrons and is reduced.

A This one is slightly more difficult, but if you work from OILRIG you'll get it. Oxidation is loss, therefore an oxidising agent aids oxidation by accepting (gaining) electrons. It is therefore A or C. Again from OILRIG, reduction is gain. Oxidising agents gain electrons, so they are therefore reduced. The answer is C.

These next two questions involve calculations that are covered more fully in chapter 5, 'Conquering calculations'.

Q Ethanol, C_2H_5OH, can be used as a fuel in some camping stoves.

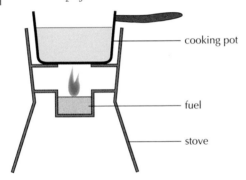

cooking pot

fuel

stove

(i) The enthalpy of combustion of ethanol given in the Data Booklet is $-1367 \text{ kJ mol}^{-1}$.

Using this value, calculate the mass of ethanol, in g, required to raise the temperature of 500 g of water from 18 °C to 100 °C.

Show your working clearly.

A This question relies on you knowing the definition for enthalpy of combustion – the energy released when 1 mole of a substance burns.

We can use the equation $E_h = cm \, \Delta T$ to work out the energy required to heat the water:

$$E_h = cm \, \Delta T$$
$$= 4 \cdot 18 \times 0 \cdot 5 \times 82$$
$$= 171 \cdot 38 \text{ kJ}$$

Our definition tells us burning 1 mole (46 g from the gram formula mass) of ethanol gives 1367 kJ of energy. It's now in proportion.

$$46 \text{ g} \rightarrow 1367 \text{ kJ}$$
$$\frac{46 \times 171 \cdot 38}{1367} \text{g} \rightarrow 171 \cdot 38 \text{ kJ}$$
$$= \underline{5 \cdot 8 \text{ g}}$$

Q Sodium benzoate is used in the food industry as a preservative. It can be made by reacting benzoic acid with a concentrated solution of sodium carbonate.

$$2C_6H_5COOH + Na_2CO_3 \rightarrow 2C_6H_5COONa + CO_2 + H_2O$$

$2C_6H_5COOH$	Na_2CO_3	$2C_6H_5COONa$	CO_2	H_2O
mass of 1 mole = 122 g	mass of 1 mole = 106 g	mass of 1 mole = 144 g	mass of 1 mole = 44 g	mass of 1 mole = 18 g

Calculate the atom economy for the production of sodium benzoate.

A Remember formulae are given in your Data Booklet.

$$\% \text{ atom economy} = \frac{\text{Mass of desired product(s)}}{\text{Total mass of reactants}} \times 100$$

This is slightly tricky. Notice you are given the mass of 1 mole of each reactant and product below the equation. However the equation tells us that 2 moles of the desired product, sodium benzoate, are produced from two moles of benzoic acid and 1 mole of sodium carbonate. When carrying out the calculation we need to multiply the mass of moles of benzoic acid and sodium benzoate by 2.

$$\% \text{ Atom economy} = \frac{2 \times 144}{(2 \times 122) + 106} \times 100$$
$$= \underline{82\%}$$

As was said at the beginning of the chapter, knowing your definitions is the key to answering many of the questions you will be asked. As you read through the following chapters of the book, hopefully you will also find that knowing your definitions will help deepen your understanding of the chemistry that has been covered in the Higher Chemistry course.

Making the most of multiple choice

This chapter covers:

- General advice
- Sometimes you just have to know it
- Use your Data Booklet to help you
- Be prepared to scribble in order to work things out
- Read the questions carefully – think about all the information
- Can you reduce the question to an either / or choice?
- When to treat the information you are given separately
- Take a moment to think about what you need to do
- Eliminate wrong answers to arrive at the correct answer

General advice

Section 1 of the Higher Chemistry paper contains fixed–response questions where you are asked to choose the correct answer from four possible options. These questions are drawn from a bank of questions, so it makes sense to practise using questions from previous papers. Working through past papers will help you become familiar with the types of questions that will be asked.

Each year there will be a balance of questions drawn from each of the topics covered in the course. The questions tend to follow the order in which the course is generally taught, that is, a few from Chemical Changes and Structure, then a few from Nature's Chemistry, followed by some from Chemistry in Society and lastly some from Researching Chemistry.

The time for both sections of the Chemistry paper is based on allowing $1\frac{1}{2}$ minutes per mark. Many candidates go through Section 1 of the paper very quickly. Take your time! Some questions will be able to be answered fairly quickly. This will create the time to allow you to think through those questions that require a bit of working out and may take more than a minute and a half to answer.

Examples

Q A compound with molecular formula $C_6H_{12}O_2$, could be

A hexanal

B hexan-2-ol

C hexan-2-one

D hexanoic acid.

A A question like this can be done very quickly if you are able to link the names of the compounds to their functional groups:

hexanal is an aldehyde and has an end carbonyl group

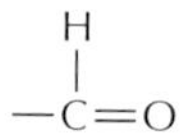

$$\begin{array}{c} H \\ | \\ -C=O \end{array}$$

hexan–2–ol is an alcohol and has a hydroxyl group

$-OH$

hexan–2–one is a ketone and has a carbonyl group in the chain

$$\begin{array}{c} O \\ \| \\ -C- \end{array}$$

hexanoic acid is a carboxylic acid and has a carboxyl group

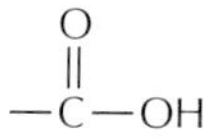

$$\begin{array}{c} O \\ \| \\ -C-OH \end{array}$$

The 2 in hexan–2–ol and hexan–2–one refers to the position of the functional group and not to the number of functional groups.

Only the carboxyl group has 2 Os, therefore the answer must be D.

Top tip

As well as marking your answer in the answer grid, circle the answer you selected on the actual question in the Section 1 question booklet. This helps when you are checking your answers at the end of the section.

Q During a redox process in acid solution, iodate ions, $IO_3^-(aq)$, are converted into iodine, $I_2(aq)$.

$$IO_3^-(aq) \rightarrow I_2(aq)$$

The numbers of $H^+(aq)$ and $H_2O(\ell)$ required to balance the ion-electron equation for the formation of 1 mol of $I_2(aq)$ are, respectively,

A 3 and 6

B 6 and 3

C 6 and 12

D 12 and 6.

A More time needs to be taken in order to answer this question. In order to get the correct answer you need to go through your rules for balancing ion–electron equations. If you try to do it in your head it is very easy to make a mistake, therefore it is much better to **write down your working** at the side of the question. This will also help when checking through your answers.

The first thing that needs to be done is balance the iodine atoms. **The common mistake is to miss this step.**

$$2IO_3^-(aq) \rightarrow I_2(aq)$$

Then balance the oxygens by adding waters.

$$2IO_3^-(aq) \rightarrow I_2(aq) + 6H_2O(\ell)$$

Six Os on the left (each iodate has three and there are two iodates) therefore $6H_2O(\ell)$ on the right.

Now balance the Hs:

$$12H^+(aq) + 2IO_3^-(aq) \rightarrow I_2(aq) + 6H_2O(\ell)$$

Twelve Hs in $6H_2O(\ell)$ on the right therefore $12H^+(aq)$ on the left.

The correct answer is D.

Top tip

- Remember, for many multiple choice questions you need to work the answer out. That might mean drawing structures or carrying out calculations.

- Remember to answer every question. There is no point in leaving a gap. If you are unsure about an answer, put down the answer you think is most likely. Remember, you can often eliminate options, giving you an easier choice.

- If you do miss a question, put a question mark sign beside the question number in the grid. This will not only help you when you are checking your answers at the end of the section, but it should prevent you putting the answer to the next question in the wrong place.

Although multiple choice items can assess direct recall of knowledge, they most often assess your understanding of the chemistry covered in the key areas defined for each part of the course as well as the knowledge and skills from Researching Chemistry.

Chemical Changes and Structure

There are three key areas in Chemical Changes and Structure:

- controlling the rate
- periodicity
- structure and bonding

Nature's Chemistry

There are seven key areas in Nature's Chemistry:

- esters, fats and oils
- proteins
- chemistry of cooking
- oxidation of food
- soaps, detergents and emulsions
- fragrances
- skin care

Chemistry in Society

There are five key areas in Chemistry in Society:

- getting the most from reactants
- equilibria
- chemical energies
- oxidising or reducing agents
- chemical analysis

Now we will look at some examples from the New Higher and Higher (Revised) papers and highlight key points when tackling multiple choice questions.

Sometimes you just have to know it

There are some multiple choice questions that will assess your direct recall of chemical information that is given in the course. There is no way for you to work these questions out. You will either know the answer or you won't!

Getting these types of questions correct is about spending time reading over your course notes or a course textbook and becoming familiar with the content.

The following question is from the key area 'Fragrances' in Nature's Chemistry.

Q   Which of the following is **not** true for an essential oil?

A   They are widely used in cleaning products.

B   They contain aroma compounds.

C   They contain volatile compounds.

D   They are water soluble.

A Notice that the word 'not' in the question is written in bold. This means that three of the statements are true for essential oils and there is one false statement. It is only by knowing the course content that you will be able to identify the false statement.

Essential oils do contain aroma compounds. They have a pleasant smell and are volatile (they evaporate easily). This is why they are used in cleaning products. They are obtained by steam distillation of plant materials. When the vapours from steam distillation are condensed, the essential oil forms a layer on top of the water layer, indicating that essential oils are not soluble in water. The correct answer is therefore option D: 'They are water soluble' is the false statement.

This next question is about understanding equilibrium from the Chemistry in Society section of the course.

Q In a reversible reaction, equilibrium is reached when

A molecules of reactants cease to change into molecules of products

B the concentrations of reactants and products are equal

C the concentrations of reactants and products are constant

D the activation energy of the forward reaction is equal to that of the reverse reaction.

A Again, this is a question where you have to consider whether each statement is true or false. This time there are three false statements and one true statement.

The question is about knowing about equilibrium at a molecular level. Even though nothing appears to be changing, at a molecular level reactant molecules are still changing to product molecules and product molecules are still reacting to the reactant particles. We describe equilibrium as a dynamic process. Option A is therefore not true.

The equilibrium position can be changed so Option B couldn't always be true.

Option D is also not true. Since there is an enthalpy change during reactions, the activation energies for the forward and back reactions will still be different.

The answer is C: the concentrations of reactants and products are constant.

This next question is about your understanding of the 'chemistry of cooking' key area in the Nature's Chemistry unit.

Q The following molecules give flavour to food.

Which of the following flavour molecules would be most likely to be retained in the food when the food is cooked in water?

A The question relies on you knowing that if hydrogen bonds can form between water molecules and flavour molecules in food then the flavour molecules will be lost into the water during cooking. The molecule that has fewest sites where hydrogen bonds can form is most likely to be retained in the food.

Option A is the correct answer as it has only one O suitable for hydrogen bond formation.

All of the other molecules have multiple sites where hydrogen bonds can form. B and C have three Os where hydrogen bonds can form to water, as well as a hydrogen of a hydroxyl group that can form a hydrogen bond to the oxygen of a water molecule. D has 2 Os and 1 N where hydrogen bonds can form. The hydrogen of the hydroxyl group can also form a hydrogen bond with the O of a water molecule.

This next question is entirely about being familiar with an experimental procedure listed in the Researching Chemistry unit of the course. Again there are three true statements and one false statement.

Q Which of the following would **not** help a student determine the end point of a titration accurately?

A swirling the flask

B using a white tile

C adding the solution dropwise near the end point

D repeating the titration.

A   Determining the end point is about judging exactly when a colour change takes place when carrying out a titration. There are three things we do to achieve this. We firstly use a white tile (fairly obvious) to enable us to see the colour clearly. When titrating we will see signs of the colour change as we approach the end point. This tends to happen in the middle of the solution where the liquid from the burette is entering the solution in the flask. When we see this happening we slow the rate at which we add the solution from the burette until we are adding it a drop at a time. Sometimes the colour will last for a few seconds then revert to the original colour. We swirl the flask to achieve thorough mixing of the solutions. It is only when we get the permanent change that we have reached the end point.

We repeat the titration to check concordancy. If we are achieving concordancy it indicates we are judging the end point accurately but repeating the titration hasn't actually helped to do that.

The correct answer is therefore D.

Use your Data Booklet to help you

For some multiple choice questions you will need to use your Data Booklet to help you answer the questions. The following questions require you to use your Data Booklet.

This question tests understanding of the electron arrangements of ions.

Q   Particles with the same electron arrangement are said to be isoelectronic. Which of the following compounds contains ions which are isoelectronic?

A Na_2S

B $MgCl_2$

C KBr

D $CaCl_2$

A   This is done simply by going to the periodic table in the Data Booklet that gives the electron arrangements for the atoms and considering the arrangements for the ions:

A Na^+ 2)8) S^{2-} 2)8)8)

B Mg^{2+} 2)8) Cl^- 2)8)8)

C K^+ 2)8)8) Br^- 2)8)18)8)

D Ca^{2+} 2)8)8) Cl^- 2)8)8)

The ions in calcium chloride are isoelectronic. The correct answer is therefore D.

The next question tests understanding of the key area 'periodicity' from Chemical Changes and Structure.

Q Which of the following atoms has least attraction for bonding electrons?

A Carbon

B Nitrogen

C Phosphorus

D Silicon

A To answer this question you need to know that an element's attraction for bonding electrons is described by its electronegativity. The higher the electronegativity of the element, the greater its attraction for bonding electrons. The atom with the least attraction for bonding electrons will have the lowest electronegativity. The simple way to answer the question is to look up the electronegativity for each of the four elements. You will find them on page 11 of your Data Booklet.

Carbon – 2·5; Nitrogen – 3·0; Phosphorus – 2·2; Silicon – 1·9

The answer is therefore D.

The answer can be found in another way.

Group 4	Group 5
6 **C** 2, 4 Carbon	7 **N** 2, 5 Nitrogen
14 **Si** 2, 8, 4 Silicon	15 **P** 2, 8, 5 Phosphorus

If you remember that electronegativity tends to increase across a period as atom size becomes smaller and that it decreases down a group as atom size becomes larger, then by thinking about the relative positions of the atoms in the periodic table you can deduce that silicon will have the lowest electronegativity. However, this would be the difficult way to answer the question.

The next question tests understanding of the key area 'oxidising and reducing agents' from the Chemistry in Society unit.

Q Which of the following will react with Br_2 but **not** with I_2?

A OH^-

B SO_3^{2-}

C Fe^{2+}

D Mn^{2+}

A To answer this question you need to use the page of your Data Booklet headed 'Electrochemical Series: Standard Reduction Potentials'.

Both Br_2 and I_2 can be reduced to ions by gaining electrons. All the species given as options appear on the right–hand side of the electrochemical series.

$$SO_4^{2-}(aq) + 2H^+(aq) + 2e^- \rightleftharpoons SO_3^{2-}(aq) + H_2O(\ell)$$

$$Cu^{2+}(aq) + 2e^- \rightleftharpoons Cu(s)$$

$$O_2(g) + 2H_2O(\ell) + 4e^- \rightleftharpoons 4OH^-(aq)$$

$$I_2(s) + 2e^- \rightleftharpoons 2I^-(aq)$$

$$Fe^{3+}(aq) + e^- \rightleftharpoons Fe^{2+}(aq)$$

$$Ag^+(aq) + e^- \rightleftharpoons Ag(s)$$

$$Hg^{2+}(aq) + 2e^- \rightleftharpoons Hg(\ell)$$

$$Br_2(\ell) + 2e^- \rightleftharpoons 2Br^-(aq)$$

$$O_2(g) + 4H^+(aq) + 4e^- \rightleftharpoons 2H_2O(\ell)$$

$$Cr_2O_7^{2-}(aq) + 14H^+(aq) + 6e^- \rightleftharpoons 2Cr^{3+}(aq) + 7H_2O(\ell)$$

$$Cl_2(g) + 2e^- \rightleftharpoons 2Cl^-(aq)$$

$$MnO_4^-(aq) + 8H^+(aq) + 5e^- \rightleftharpoons Mn^{2+}(aq) + 4H_2O(\ell)$$

When we look at the table of electrode potentials we see that:

- Two of the species given as options, SO_3^{2-} and OH^-, appear above the equations involving iodine and the bromine. They will behave in the same way with both and are in fact strong enough reducing agents to be able to reduce bromine and iodine.
- One species, Mn^{2+}, lies below both equations – Mn^{2+} will also behave in the same way with both bromine and iodine. It isn't a strong enough reducing agent to be able to react with either.
- One species, Fe^{2+}, lies between the equations. We would therefore expect it to behave differently with bromine and iodine. It is a strong enough reducing agent to reduce bromine but will not be able to reduce iodine.

The answer is therefore C.

Here is another similar question that tests understanding of the same key area.

Q Which of the following ions could be used to oxidise iodide ions to iodine?

$$2I^-(aq) \rightarrow I_2(s) + 2e^-$$

A $SO_4^{2-}(aq)$

B $SO_3^{2-}(aq)$

C $Cr^{3+}(aq)$

D $Cr_2O_7^{2-}(aq)$

A The equations involving the species appear in the following order in the electrochemical series.

$$SO_4^{2-}(aq) + 2H^+(aq) + 2e^- \rightleftharpoons SO_3^{2-}(aq) + H_2O(\ell)$$

$$I_2(s) + 2e^- \rightleftharpoons 2I^-(aq)$$

$$Cr_2O_7^{2-}(aq) + 14H^+(aq) + 6e^- \rightleftharpoons 2Cr^{3+}(aq) + 7H_2O(\ell)$$

To oxidise iodide ions to iodine, the species must act as an oxidising agent, it must be able to be reduced. The two ions given as options that can be reduced are $SO_4^{2-}(aq)$ and $Cr_2O_7^{2-}(aq)$. Both species appear on the left-hand side of the electrochemical series. The lower on the left of the series, the stronger the oxidising agent. $Cr_2O_7^{2-}(aq)$ is therefore the stronger oxidising agent and will be able to oxidise the iodide ions. The answer is therefore D.

Be prepared to scribble in order to work things out

Sometimes you will need to carry out simple calculations to arrive at the correct answer.

This next question tests understanding of Hess's Law from the key area 'chemical energy' in Chemistry in Society.

Q Consider the reaction pathways shown below.

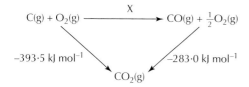

According to Hess's law, the enthalpy change for reaction **X** is

A $-676 \cdot 5 \, \text{kJ mol}^{-1}$

B $-110 \cdot 5 \, \text{kJ mol}^{-1}$

C $+110 \cdot 5 \, \text{kJ mol}^{-1}$

D $+676 \cdot 5 \, \text{kJ mol}^{-1}$

A Hess's Law states that the energy change for a chemical change is the same no matter the route taken. Therefore the energy change X for

$$C(g) + O_2(g) \longrightarrow CO(g) + \tfrac{1}{2}O_2(g)$$

must be the same as going from

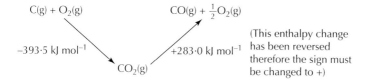

(This enthalpy change has been reversed therefore the sign must be changed to +)

The answer is therefore B ($-393 \cdot 5 \, \text{kJ mol}^{-1}$ + $+283 \cdot 0 \, \text{kJ mol}^{-1}$ = $-110 \cdot 5 \, \text{kJ mol}^{-1}$)

Be prepared to scribble on your question paper in order to work this out.

Sometimes you might be required to do quite a bit of working. This next question appeared in the 2015 Higher (Revised) paper, but a similar question could easily appear in a New Higher paper.

Q 4·6 g of sodium is added to 4·8 litres of oxygen to form sodium oxide.

When the reaction is complete, which of the following statements will be true?

(Take the volume of 1 mole of oxygen to be 24 litres.)

A 0·10 mol of oxygen will be left unreacted.

B 0·10 mol of sodium will be left unreacted.

C 0·15 mol of oxygen will be left unreacted.

D 0·20 mol of sodium oxide will be formed.

A Where do you start? You can't do this question without first writing the equation for the reaction:

$$4Na(s) + O_2(g) \rightarrow 2Na_2O(s)$$

You then have to work out the quantities of reactants you have, the quantities that will react and the quantity of product produced. There is no shortcut to getting the correct answer.

The working for the calculations you would need to do has been set out for you. You need to be prepared to scribble this working on your question paper or spare paper that is given to you for doing working on. Examining the calculation and the options given in the question you will see that the answer must be C.

4·6 g sodium is equal to 0·2 moles $\left(\dfrac{4·6}{23} \right)$; 4·8 litres of oxygen is also 0·2 moles $\dfrac{4·8}{24}$

$$4Na(s) + O_2(g) \rightarrow 2Na_2O(s)$$

Initial moles	0·2	0·2	
Moles reacting / produced	0·2	0·05	0·1
Moles remaining	0	0·15	0·1

Examining the calculation and the options given you will see that the correct answer is C.

You might suspect the answer is either A or C since both statements involve unreacted oxygen. If you had to guess at an answer the smart guess might be to guess one of these two options.

Read the questions carefully – think about all the information

Q Which of the following mixtures will form when NaOH(aq) is added to a mixture of propanol and ethanoic acid?

A Propanol and sodium ethanoate

B Ethanoic acid and sodium propanoate

C Sodium hydroxide and propyl ethanoate

D Sodium hydroxide and ethyl propanoate.

A You might read this question quickly and see propanol and ethanoic acid. You immediately think 'This is a question about esters' and the answer must be C or D and then work out it must be C since propanol and ethanoic acid make the ester, propyl ethanoate. You would have just made a very common mistake, and this is why you need to read the question carefully!

There is another piece of information in the stem that needs to be considered.

What is added to an alcohol / carboxylic acid mixture to bring about esterification? A few drops of concentrated sulfuric acid. But that's not what is being added. Sodium hydroxide solution, an alkali, is being added. Alkalis react with acids to give salts. The sodium hydroxide solution will react with the ethanoic acid and give the salt sodium ethanoate. It won't react with propanol.

The answer is therefore A.

Can you reduce the question to an either/or choice?

This is particularly true when a list of numerical values is given.

Q Oils contain carbon-to-carbon double bonds which can undergo addition reactions with iodine.

The iodine number of an oil is the mass of iodine in grams that will react with 100 g of oil.

Which line in the table shows the oil that is likely to have the lowest melting point?

	Oil	Iodine number
A	Corn	123
B	Linseed	179
C	Olive	81
D	Soya	130

A Logically when looking at a question like this it is likely to be the one with either the highest or lowest number, in this case iodine number.

Looking at the information in the stem, the more carbon-to-carbon double bonds, the more iodine can react with the oil. Therefore the oil with the highest iodine number has most double bonds and is the most unsaturated.

You now need to think about melting points. The more unsaturated the oil, the lower the melting point will be.

The answer is therefore B, linseed oil.

More double bonds means the oil molecules can't pack as closely, and will have weaker intermolecular forces leading to lower melting points.

When to treat the information you are given separately

In this next question you are given two separate pieces of information to consider. Considering the first piece will narrow your choice to two options. Considering the second piece leads you to the correct answer.

Q Carvone is a natural product that can be extracted from orange peel.

Carvone

Which line in the table correctly describes the reaction of carvone with bromine solution and with acidified potassium dichromate solution?

	Reaction with bromine solution	Reaction with acidified potassium dichromate solution
A	no reaction	no reaction
B	no reaction	orange to green
C	decolourises	orange to green
D	decolourises	no reaction

A The first piece of information is about reaction with bromine solution. Bromine solution is decolourised when added to compounds containing double carbon-to-carbon bonds. The structure shows that carvone contains a double carbon-to-carbon bond. Bromine solution therefore decolourises when added to carvone. This therefore reduces the options to be considered to C and D.

Carvone contains a carbonyl group. Will it react with acidified potassium dichromate? Acidified potassium dichromate is used to distinguish between aldehydes and ketones. It will oxidise aldehydes to carboxylic acids but does not react with ketones. Carvone is a ketone – the carbon of the carbonyl group is attached to two other carbons. It will not therefore react with the acidified potassium dichromate solution. The correct answer is therefore D.

It doesn't matter which piece of information you consider first. Thinking about the dichromate first would have reduced the options to A and D. Considering bromine would then have led to the same answer.

Take a moment to think what you need to do

In this next question you might launch into doing something you don't need to do.

Q Which of the following is an isomer of 2,2-dimethylpentan-1-ol?

A $CH_3CH_2CH_2CH(CH_3)CH_2OH$

B $(CH_3)_3CCH(CH_3)CH_2OH$

C $CH_3CH_2CH_2CH_2CH_2CH_2CH_2CH_2OH$

D $(CH_3)_2CHC(CH_3)_2CH_2CH_2OH$

A The stem of the question gives the name of an alcohol and asks you to select the correct option from some shortened structural formulae.

You might be tempted to start drawing full structural formulae, but do you need to do this?

From the name you can work out that 2,2-dimethylpentan-1-ol has 7 carbons: 5 from the main chain and 2 from the two methyl groups. You simply need to look for the option with 7 carbons. Option A has 6, option B has 7 and options C and D have 8.

The correct answer is therefore B.

But what if two options had 7 carbons? Then you would have to consider numbers of Hs and Os in the formulae. You might have to draw out 2,2-dimethylpentan-1-ol.

Eliminate wrong answers to arrive at the correct answer

In this next question three answers can be eliminated, leaving the correct answer.

Q The following equilibrium exists in bromine water.

$$\underset{(red)}{Br_2(aq)} \ + \ H_2O(\ell) \ \rightleftharpoons \ \underset{(colourless)}{Br^-(aq)} \ + \ 2H^+(aq) \ + \ \underset{(colourless)}{OBr^-(aq)}$$

The red colour of bromine water would fade on adding a few drops of a concentrated solution of

A HCl

B KBr

C AgNO$_3$

D NaOBr

A For the red colour of the bromine to fade, the equilibrium position needs to move to the right.

By considering each option in turn we can start to eliminate wrong answers.

Consider Option A: Adding HCl increases the concentration of H$^+$(aq). H$^+$(aq) appear on the right of the equation, therefore increasing their concentration will push the equilibrium to the left. The red colour will become more intense. It's not A.

Consider Option B: Adding KBr increases the concentration of Br$^-$(aq). Br$^-$(aq) appear on the right of the equation, therefore increasing their concentration will push the equilibrium to the left as well. Again the red colour will become more intense. It's not B.

Consider Option C: Ag$^+$(aq) and NO$_3^-$(aq) don't appear in the equation. You think 'I'm not sure. What about Option D?'.

Again with D you are increasing a species that appears on the right of the equation (OBr⁻(aq)) and this will push the equilibrium to the left, so C must be the correct answer.

What's the chemistry behind C being the correct answer?

When you add silver nitrate, the silver ions react with bromide ions to give silver bromide which is insoluble:

$$Ag^+(aq) + Br^-(aq) \rightarrow AgBr(s)$$

The concentration of bromide ions is decreased. The equilibrium will move to the right causing the red colour of the bromine to fade.

Tidying up trends in the periodic table

This chapter covers:

- Periodic properties
- Trends across a period
- Trends down a group

Periodic properties

It took chemists a long time to gain an understanding of the relationships between elements.

Dmitri Mendeleev, a Russian chemist, was the scientist who finally suggested a system for classifying the elements. His system brought insight to many of his fellow scientists who had struggled to understand how elements related to each other. This system became the basis for the modern periodic table.

Just as for those early pioneers of chemical understanding, knowledge of the periodic table can bring great insight to us as well.

The periodic table is based on patterns. When Mendeleev first arranged the elements that were known at that time he likened it to the card game *Patience*, where the objective is to arrange the cards in suits according to their numerical order.

The modern periodic table is based on electron arrangements of the elements, with the periods corresponding to the filling of electron shells and the main groups being atoms with the same number of outer shell electrons. These arrangements lead to patterns as we go across a period and also as we go down groups. These patterns can explain the properties of the elements.

When thinking about periodic properties there are key features we need to understand.

There are three periodic properties that you may be asked to explain or understand:

- atom size (covalent radius)
- ionisation energy
- electronegativity

These are best understood by thinking about a period and a group separately.

Trends across a period

The second period corresponds to the filling of the second electron shell and contains elements from lithium, with 3 protons in its nucleus, to neon, with 10 protons in its nucleus.

3	4	5	6	7	8	9	10
Li	**Be**	**B**	**C**	**N**	**O**	**F**	**Ne**
2, 1	2, 2	2, 3	2, 4	2, 5	2, 6	2, 7	2, 8
Lithium	Beryllium	Boron	Carbon	Nitrogen	Oxygen	Fluorine	Neon

For each successive element across the period one more proton is added to the nucleus and one more electron is added to the outer shell.

The really important factor is the extra proton being added to the nucleus.

As we go across the period the number of electron shells stays the same but the number of protons in the nucleus is increasing. The pull by the nucleus on the outer electrons is getting stronger. The outer shell electrons are therefore pulled closer to the nucleus and are more tightly held.

The effects on atom size and first ionisation energy are as follows.

- Atom size decreases across a period. (Notice there is no covalent radius quoted in the Data Booklet for neon as atoms of neon do not form bonds to each other.)
- First ionisation energy increases. It's harder to pull electrons away.

Electronegativity – the smaller the size of an atom, the greater the pull will be on bonded electrons – will increase across the period.

Trends down a group

As we go down a group we see that all of the atoms in the group have the same number of electrons in the outer shell. The number of protons in the nucleus increases, but the number of electron shells between the nucleus and the outer electrons also increases. The inner electron shells, those between the nucleus and the outer electron shell, shield the outer electrons from the nuclear pull. Distance from the nucleus and shielding by inner electron shells are the dominant factors when considering properties down a group.

Down a group we see:

- atom size increases due to an extra shell of electrons being added to each successive element
- shielding of the nuclear pull by inner electron shells increases
- nuclear attraction for outer shell electrons decreases.

It will be easier therefore to remove an electron from an atom as we go down a group. Ionisation energy will decrease down the group as will electronegativity.

Thinking about how nuclear charge, size of atoms and shielding effects by inner electron shells affect properties will allow us to tackle the questions that examine our understanding of periodic trends.

| 3 |
| Li |
| 2, 1 |
| Lithium |
| 11 |
| Na |
| 2, 8 , 1 |
| Sodium |
| 19 |
| K |
| 2, 8, 8, 1 |
| Potassium |
| 37 |
| Rb |
| 2, 8, 18, 8, 1 |
| Rubidium |
| 55 |
| Cs |
| 2, 8, 18, 18, 8, 1 |
| Caesium |
| 87 |
| Fr |
| 2, 8, 18, 32, 18, 8, 1 |
| Francium |

Examples

Q Which entry in the table shows the trends in the electronegativity values of the elements in the periodic table?

	Across a period	Down a group
A	decrease	decrease
B	decrease	increase
C	increase	decrease
D	increase	increase

A This question is a straightforward recall of what you have been taught. The answer is C, electronegativities increase across a period and decrease down a group.

Top tip

Unsure of the answer? Check the Data Booklet. Remember, electronegativities are given in a table.

Q Which of the following elements has the greatest attraction for bonding electrons?

A Lithium

B Chlorine

C Sodium

D Bromine

A Attraction for bonding electrons is expressed as electronegativity. The simplest way to answer the question is to go to the Data Booklet and look up the electronegativities of each element. Chlorine with electronegativity 3·0 has the highest electronegativity and therefore the greatest attraction for bonding electrons.

If you remember that electronegativities increase across a period as atom size decreases, and decrease down groups as atom size increases, then you can deduce that given two elements from Group 1 and two from Group 7, the element with most attraction will be the one in Group 7 that has the smaller atom size (that is, chlorine).

The answer is B.

Q The covalent radius is a measure of the size of an atom.

(i) Explain why covalent radius decreases across the period from sodium to chlorine.

A This explanation was for 1 mark. What's happening as we go across the period? For each successive element one more proton is being added to the nucleus and one more electron is being added to the outer shell.

Explanation
Because the nuclear charge is increasing as we go across the period there is a greater pull on the outer electron shell, therefore the size of atoms will decrease across the period.

Q (a) An alternative to common salt contains potassium ions and chloride ions.

(i) Write an ion-electron equation for the first ionisation energy of potassium.

A This is a straightforward question, particularly if you look at the table of ionisation energies in the Data Booklet.

The notes at the top give you the form of the equation you want.

Notes: The first ionisation energy for an element E refers to the reaction $E(g) \rightarrow E^+(g) + e^-$; the second ionisation energy refers to $E^+(g) \rightarrow E^{2+}(g) + e^-$; etc.

You simply need to substitute E with the symbol for potassium.

Answer:

$$K(g) \rightarrow K^+(g) + e^-$$

The question goes on to ask you to give an explanation relating to first ionisation energy.

Q **Explain clearly** why the first ionisation energy of potassium is smaller than that of chlorine.

A This was worth 3 marks. '**Explain clearly**' in bold is a big hint that a detailed explanation is required.

Where do you start?

This is where you need to start by writing down some information you know:

K   19 protons   electron arrangement 2)8)8)1

Cl   17 protons   electron arrangement 2)8)7

As soon as you do this you should be able to see the factors responsible for the differences.

Although potassium has more protons, it has a smaller first ionisation energy. This is because it has four electron shells compared to the three of chlorine. The outer electron shell of potassium will be further from the nucleus and will be shielded from the nuclear pull by more inner electron shell.

Writing down the information you know or can find from the Data Booklet and commenting on it can give you a good start in any question of this type.

Q Explain fully why the second ionisation energy is much greater than the first ionisation energy for Group 1 elements.

A Again this is another question where it may help to write electron arrangements down.

The second ionisation energy is the energy to remove a second mole of electrons. Again, the form of the equation is given at the top of the table of ionisation energies.

For sodium:

$$Na(g) \rightarrow Na^+(g) + e^-$$

Electron Arrangements   2)8)1   2)8)

$$Na^+(g) \rightarrow Na^{2+}(g) + e^-$$

Electron Arrangements   2)8)   2)7

Explanation

The second ionisation energy is much greater because the electron being removed is being removed from a shell that is closer to the nucleus. There is a greater nuclear pull on this electron since it is on a shell closer to the nucleus and there is one less shell shielding the electron from the nuclear pull.

> **Top tip**
>
> There is always a large jump in ionisation energy when the electron to be removed has to be removed from a shell closer to the nucleus.

This last question is taken from the 2006 Higher and illustrates this point graphically.

Q The spike graph shows the variation in successive ionisation energies of an element, **Z**.

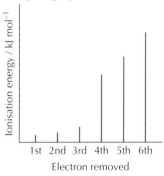

In which group of the periodic table is element **Z**?

A 1

B 3

C 4

D 6

A The large jump comes with the fourth ionisation energy. The element has, therefore, three electrons in its outer shell with the fourth electron having to be removed from a shell closer to the nucleus.

The answer is therefore B.

Becoming better at bonding

This chapter covers:

- Bonding and structure
  - o The bonding continuum
  - o Bonding and structure in elements and compounds
  - o Polar and non-polar molecules
- Intermolecular forces
  - o How do the forces arise?

Your understanding of how substances hold together would have begun earlier in your chemistry learning experience but will have been developed further at Higher. At Higher your understanding should develop from a basic understanding of the types of bonds in elements and compounds to a more complete understanding of the bonding and structures of elements and compounds. You will also have been introduced to the forces that exist between molecules in molecular substances.

This understanding of the bonding and structure of substances helps us to make sense of many of the physical properties such as melting point, boiling point and solubility.

There are always a number of questions in the course exam that will test your understanding of the structures and properties of substances in terms of the bonding and intermolecular forces within the substances.

Firstly let's summarise what you should know.

Bonding and structure

The bonding continuum

At Higher we develop the idea that bonding isn't divided into neat groups but rather there is a bonding continuum with compounds having covalent bonds at one end and compounds with ionic bonding at the other. Polar covalent bonds lie somewhere between the two.

We have already seen in the last chapter that elements have different electronegativities. These can be used to explain why some compounds are ionic and some are covalent.

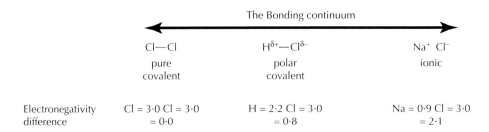

The Bonding continuum		
Cl—Cl	$H^{\delta+}$—$Cl^{\delta-}$	Na^+ Cl^-
pure covalent	polar covalent	ionic

Electronegativity difference	Cl = 3·0 Cl = 3·0 = 0·0	H = 2·2 Cl = 3·0 = 0·8	Na = 0·9 Cl = 3·0 = 2·1

When the bonding atoms have the same electronegativity, pure covalent bonds form. A covalent bond is a shared pair of electrons. The atoms are held together by the attraction of the nuclei for the shared pair. The electrons become localised between the atom centres.

Note this is not the case with metal elements. Metals have low electronegativity meaning that the forces of attraction for the bonding electrons are so low that the electrons cannot become localised between the atom centres. The electrons are said to be delocalised. This explains the electrical conductivity of metals.

In a polar covalent bond the electrons are pulled towards the atom with the greater electronegativity. The symbols δ+ and δ- are often used to show the polarity of the bonds.

When the bonding atoms have a large electronegativity difference there will be electron transfer and ions will form.

Bonding and structure in elements and compounds

The tables below describe what you should know in terms of bonding and structure.

Elements			
Metals			
Example	Bonding	Structure	Properties
Na, Cu, Sn	metallic	lattice	Conduct electricity when solid or liquid.
Non-metals			
Example	Bonding	Structure	Properties
C, Si, B	covalent	network	Do **not** conduct electricity in any state. Have very high melting points.

Diamond

Non-metals continued			
Examples	Bonding	Structure	Properties
O_2, N_2, S	covalent	molecular	Do **not** conduct electricity in any state. Low melting and boiling points.

$N\equiv N$ S_8 Crown

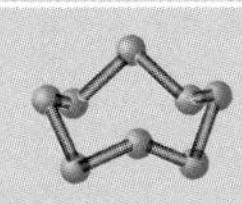

| He, Ne, Ar | – | monatomic | Very low melting and boiling points. |

Compounds			
Ionic			
Examples	Bonding	Structure	Properties
Na^+Cl^- $(K^+)_2O^{2-}$ $Ca^{2+}SO_4^{2-}$	ionic	lattice	Conduct electricity when molten or in solution. Solid at room temperature. Tend to have high melting points. Some soluble in water (always refer to Data Booklet).

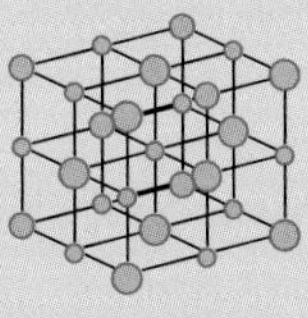

Na^+Cl^- lattice

Covalent			

A. Many covalent compounds exist as discrete molecules. The atoms within the molecule are covalently bonded to each other.

Examples	Bonding	Structure	Properties
CO_2, CH_4, NH_3, H_2O, $C_6H_{12}O_6$	covalent	molecular	Do **not** conduct electricity in any state. Usually liquids or gases at room temperature.

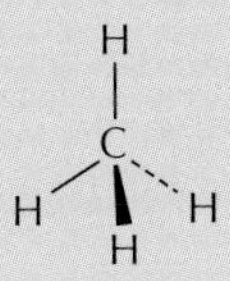

CH_4

$O=C=O$
CO_2

B. Some covalent compounds have a covalent network structure. The atoms are covalently bonded within the network structure.

Examples	Bonding	Structure	Properties
SiO_2, SiC	covalent	network	Do **not** conduct electricity in any state. Have very high melting points. Unreactive. Insoluble in water.

SiC

Polar and non-polar molecules

The presence of polar covalent bonds and the shapes of molecules can lead to molecules being polar, that is, having a permanent dipole with a positive end and a negative end.

Chloromethane is a simple example of a polar molecule. Electrons are pulled to the chlorine end of the molecule due to chlorine's high electronegativity, making this end slightly negative.

The polarity of the bonds and the shapes of the molecule affect the strength of intermolecular forces.

Intermolecular forces

The forces that exist between molecules are referred to as van der Waals' forces. There are two types: London dispersion forces and permanent dipole–permanent dipole interactions. Hydrogen bonds are a particularly strong type of permanent dipole–permanent dipole interaction.

How do the forces arise?

London dispersion forces

The electrons surrounding atoms and molecules are moving at high speed. The electrons can be unevenly distributed around the particle. Momentarily one part of the molecule may be slightly more negative. We say that the particle has a temporary dipole and that this induces a dipole on a neighbouring particle, creating a force of attraction. This force of attraction is referred to as a London dispersion force.

London dispersion forces due to temporary dipoles caused
by electron movement.

The positions of London dispersion forces are constantly shifting as the electrons move. The more electrons a molecule has, the larger the variations in electrical density can be, therefore the stronger the London dispersion forces will be.

Permanent dipole–permanent dipole interactions

Polar molecules have permanent dipoles. This means that the positive end of one molecule will be attracted to the negative end of another molecule. The bonds in hydrogen chloride are polar due to the chlorine atom having a larger electronegativity than hydrogen. This leads to permanent dipole–permanent dipole interactions occurring between the molecules.

$\delta+$ $\delta-$ $\delta+$ $\delta-$ $\delta+$ $\delta-$ ·····	**permanent dipole-permanent**
H—Cl ········· H—Cl ········· H—Cl	**dipole interaction**

Permanent dipole–permanent dipole interactions are stronger than London dispersion forces.

Hydrogen bonds

In molecules containing a hydrogen atom bonded to a nitrogen atom or an oxygen atom or in the compound hydrogen fluoride, the permanent dipole–permanent dipole interactions are so strong that these intermolecular forces are given a special name and referred to as hydrogen bonds.

A hydrogen bond between
water molecules.

Ammonia and water have much higher melting and boiling points than might be predicted from the melting points of the other hydrides in their groups. This is due to the presence of hydrogen bonds between molecules.

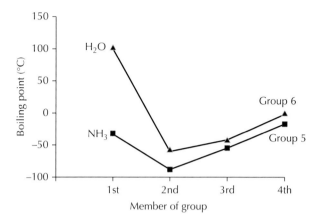

Melting and boiling point of substances depend upon the type and numbers of intermolecular forces occurring between molecules. The stronger the intermolecular forces the more energy is required to break them.

Summary

Type of molecule	Strongest type of van der Waals' force	Strength	Example
Non-polar	London dispersion force	I N C R E A S I N G	
Polar	Permanent dipole–permanent dipole interaction		
Polar containing hydrogen bonded to N, O or F	Hydrogen bond		

Let's look at some examples from past papers.

This first question is a straight recall of knowledge.

Q Which line in the table is correct for the polar covalent bond in hydrogen chloride?

	Relative position of bonding electrons	Dipole notation
A	H —•• Cl	$\overset{\delta+}{H} — \overset{\delta-}{Cl}$
B	H ••— Cl	$\overset{\delta+}{H} — \overset{\delta-}{Cl}$
C	H —•• Cl	$\overset{\delta-}{H} — \overset{\delta+}{Cl}$
D	H ••— Cl	$\overset{\delta-}{H} — \overset{\delta+}{Cl}$

A Answer A is correct.

Explanation
Chlorine is more electronegative than hydrogen (Cl electronegativity 3·0; H electronegativity 2·2). The electrons in the bond are pulled to the chlorine end of the bond and this results in the chlorine having a slightly negative charge ($\delta-$) and the hydrogen having a slightly positive charge ($\delta+$).

Q Which of the following has more than one type of van der Waals' force operating between its molecules in the liquid state?

A   Br—Br

B   O=C=O

C   (structure: N bonded to H, H, H)

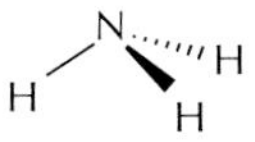

D   (structure: C bonded to H, H, H, H)

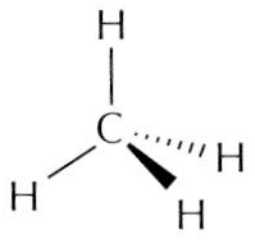

A Answer C is correct.

Explanation
This question highlights a factor many candidates forget. With polar molecules, more than one type of intermolecular force exists between molecules in the liquid state. The

strongest intermolecular force between ammonia molecules will be hydrogen bonds but there will also be London dispersion forces present between molecules as well. This is perhaps better understood by thinking about a molecule such as butan-1-ol.

Both hydrogen bonds and London dispersion forces are present in butan-1-ol.

Hydrogen bonds can form between adjacent hydroxyl groups. Weaker London dispersion forces can form between the hydrocarbon chains of adjacent molecules.

Now we come to some questions from the written section of the paper.

Q Arsenic(III) sulfide is an orange-yellow powder which is insoluble in water. Below 310 °C it can sublime, turning from a solid to a gas.

Name the type of bonding and structure present in arsenic(III) sulfide.

Top tip

If asked to state the type of bonding in a compound, always work from the properties and not from the name.

A A question like this crops up from time to time. It tests your understanding that bonding in compounds is classified according to properties rather than based on the name of the compound. Arsenic is a metal so you might think that arsenic(III) sulphide should be ionic.

The properties say otherwise! The key is that it sublimes. This means it changes from solid to gas without becoming liquid. As it can form a gas it must be made of molecules in the gas phase. The melting point is fairly low, which indicates the solid is also molecular rather than a network. A much higher temperature would be required to break the strong covalent bonds in a network. The answer is therefore covalent molecular.

Q Liquid hydrogen sulfide has a boiling point of –60 °C.

Name the strongest type of intermolecular force present in liquid hydrogen sulfide and state how this force arises.

A This is a simple one mark question. However, seeing the hydrogens, might you be tempted to answer 'hydrogen bonding'? Or seeing the shape, do you immediately think permanent dipole–permanent dipole attractions?

It's not hydrogen bonding as the hydrogen is not attached to N, O or F. The low boiling point also suggests it isn't hydrogen bonding. So it is either London dispersion forces or permanent dipole–permanent dipole interactions. Sulfur has electronegativity 2·5 and hydrogen has electronegativity 2·2. The answer will be permanent dipole–permanent dipole interactions.

This question focuses in on electronegativity.

Q Explain the difference in polarities of ammonia and trichloramine molecules.

ammonia   trichloramine

A The first thing you must remember to do in your answer is state the polarity of each molecule and then explain your reasoning.

Ammonia is a polar molecule because nitrogen has an electronegativity of 3·0 and hydrogen has an electronegativity of 2·2. Electrons are pulled to the nitrogen end of the molecule making that slightly negative. In trichloramine the N and Cl atoms both have an electronegativity of 3·0. The electrons in the bonds will be shared equally, meaning the molecule should be non-polar.

This next question is worth three marks so a very full explanation is required. However, it gives you guidance about the points to include in your answer.

Q The melting point of sulfur is much higher than that of phosphorus.

Explain fully, in terms of the structures of sulfur and phosphorus molecules and the intermolecular forces between molecules of each element, why the melting point of sulfur is much higher than that of phosphorus.

A Structure your answer according to the question. Firstly, give the structures of sulfur and phosphorus molecules.

Sulfur S_8 molecules
Phosphorus P_4 molecules

You can draw them if you want, but you don't have to.

Secondly, state the type of intermolecular forces existing between molecules of the elements.

There are London dispersion forces between the molecules of both.

Thirdly, think about relative size of forces. S_8 molecules will have more electrons than P_4 molecules, therefore they will have stronger / more London dispersion forces between molecules. Hence sulfur will have a higher melting point.

Q The boiling point of hydrogen fluoride, HF, is much higher than the boiling point of F_2.

H—F   F—F
boiling point: 19·5 °C   boiling point: −188 °C

Explain fully why the boiling point of hydrogen fluoride is much higher than the boiling point of fluorine.

In your answer you should mention the intermolecular forces involved and how they arise.

A Again this is a question looking at intermolecular forces.

Explanation
Strong hydrogen bonds exist between molecules of hydrogen fluoride due to the large electronegativity difference between hydrogen and fluorine. Only weak London dispersion forces due to temporary dipoles caused by the movement of electrons exist between molecules of fluorine.

London dispersion forces are much weaker than hydrogen bonds. Less energy is required to break the weak London dispersion forces, therefore the boiling point of fluorine will be much lower than that of hydrogen fluoride.

This next question gives you plenty of opportunity to give the correct answer and lots of opportunities to get it wrong.

Q On the diagram below use a dotted line to show **one** hydrogen bond that could be made between a behenic acid molecule and the keratin.

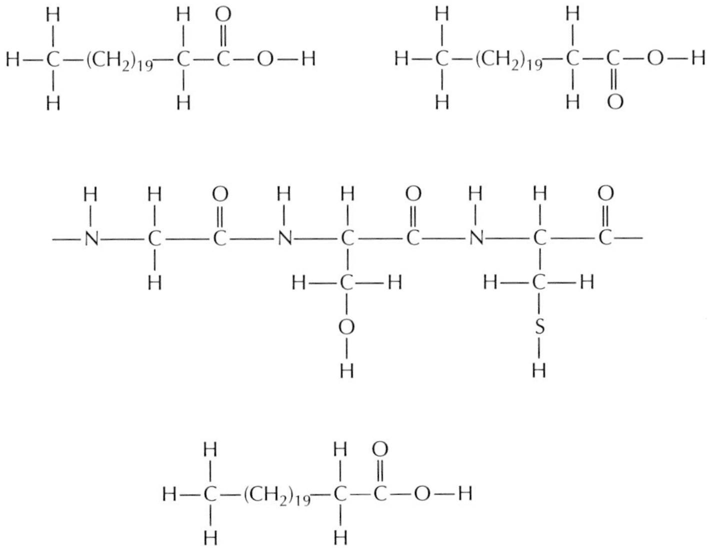

A The dotted lines indicate the six most obvious hydrogen bonds that could form.

Note that the hydrogen bonds form between a hydrogen attached to an oxygen or a nitrogen, and a nitrogen or oxygen on an adjacent molecule.

Conquering calculations

This chapter covers:

- Important points to remember when tackling calculation questions
- Calculations that test general numeracy skills
- Calculations that are taught as part of the course

Calculations make up between around 20–25% of the marks that are available in the Higher Chemistry written paper. This means that most of the calculation types you will cover during the course are likely to appear in the paper. This is very different from the situation in National 5 exams where fewer marks are allocated to calculations.

It is therefore important if you are to achieve a good grade in Higher Chemistry that you spend time conquering calculations. Teachers often hear students say 'I just can't do calculations' and there is no doubt that they can be daunting. However, following a few simple rules can give you the confidence to conquer calculations.

There are two distinct types of calculations in the Higher Chemistry paper that we will deal with in this chapter:

- calculations that test general numeracy skills and
- specific chemistry calculations that are taught as part of the course.

First, here are some important general points to remember when tackling calculations.

Important points to remember when tackling calculation questions

Know your calculator

You will need to use a calculator for some of the questions in your exam, so it is important that you are familiar with the various functions of your calculator. Check your calculator is working before you go into an exam. The last thing you want is to find it is not working and you need to borrow an unfamiliar calculator.

Always show your working clearly

If you give the correct numerical value for an answer you will be awarded full marks for the question (unless the question states 'You must show your working'). However, if you give a wrong answer, the marker can still award you partial marks either because you have calculated some part of the question correctly or the marker can see in your working evidence that you understand a concept that you have been taught as part of the course.

Check your units

There is one mark available for correct units in the paper.

In most questions the units will be stated in the stem of the question. For example, a question may start 'Calculate the enthalpy change, in kJ mol^{-1}, for ...'. There is therefore no need for you to give a unit with your answer, although you can do so if you wish. However, if you give a wrong unit your answer would be deemed to be incorrect and you could not be awarded the mark.

There will be one question where the units are not given and you will be awarded a mark if you give the correct unit.

Top tip

When giving your final answer, check the stem of the question to see if units were given and make sure the units you give are the same. If no units are given in the stem, think about the units you give in your answer. For example, if you have been asked for the volume of gas released in a reaction, you need to decide if the answer you have worked out should be in litres or cubic centimetres.

Check your answer is reasonable

We need to apply some good old-fashioned common sense. Think about the amount of material you started with and consider what would be a reasonable amount in your answer. If you were making an ester using 3·5 g of ethanol, are you going to end up with 594 g of ester? The answer is obviously 'No'. It is therefore likely that you have made a mistake somewhere along the line and the answer should really be 5·94 g. Just stopping and thinking for a moment can avoid making silly mistakes.

Calculations that test general numeracy skills

About a third of the marks allocated to calculations are allocated to questions that test general numeracy skills. It is a benefit to anyone who wishes to go on and work or study in science-related areas to have good basic numeracy skills. The questions testing these skills simply use a chemical context to test whether you have a grasp of basic arithmetical operations such as percentages, proportion, etc. There would be no need for you to have studied the Higher Chemistry course to be able to tackle them.

Let's look at a couple of examples.

Q Theobromine, a compound present in chocolate, can cause illness in dogs and cats.

To decide if treatment is necessary, vets must calculate the mass of theobromine consumed.

1·0 g of chocolate contains 1·4 mg of theobromine.

Calculate the mass, in mg, of theobromine in a 17 g biscuit of which 28% is chocolate.

Show your working clearly.

A This question can be done by anyone who is competent working with percentages and proportion. You don't have to have studied Chemistry to do it.

Also notice the units, mg, are given in the stem of the question.

28% of the biscuit is chocolate and the biscuit weighs 17 g.

Using percentages:

The weight of chocolate in the biscuit is therefore $17 \times \dfrac{28}{100} g$

$$= 4 \cdot 76 \ g$$

Using proportion:

1·0 g chocolate contains 1·4 mg theobromine

4·76 g contains $4 \cdot 76 \times 1 \cdot 4$ mg theobromine

$$= 6 \cdot 664 \text{ (the units are not required as they are given in the stem)}$$

What about significant figures? Significant figures help show how accurate you should be with your result. The numbers used in the stem of the question (1·0; 17; 28) all have two significant figures.

The answer should therefore be rounded to 6·7 (mg). You should really round an answer to the lowest number of significant figures given in the numbers in the question.

However it is unlikely that you would be penalised for leaving your answer as 6·664. The two marks in the question would be for showing you could use percentages and proportion with no additional credit for rounding.

Here is another question that is simply a numeracy calculation set in a chemical context.

Q The level of hypochlorite in swimming pools needs to be maintained between 1 and 3 parts per million (1–3 ppm).

400 cm^3 of a commercial hypochlorite solution will raise the hypochlorite level of 45 000 litres of water by 1 ppm.

Calculate the volume of hypochlorite solution that will need to be added to an Olympic-sized swimming pool, capacity 2 500 000 litres, to raise the hypochlorite level from 1 ppm to 3 ppm.

A The question might at first look daunting but it is simply a matter of using proportion. The question may include unfamiliar units: parts per million, but this shouldn't put you off.

You want to raise the hypochlorite level in the swimming pool by 2 ppm.

If 400 cm^3 of hypochlorite solution is needed to raise 45 000 litres of water by 1 ppm, how much is needed to raise it by 2? Easy! Just multiply by 2: the answer is 800 cm^3.

However, you want to do this for an Olympic-sized pool, so you need to use proportion.

Setting it out:

45 000 litres	by 1 ppm	⟷	400 cm^3
45 000 litres	by 2 ppm	⟷	2×400 cm^3
2 500 000 litres	by 2 ppm	⟷	$2 \times 400 \times \dfrac{2\,500\,000}{45\,000}$
		=	44 444 cm^3

Notice two things:

1. Significant figures: 800 has three significant figures so the answer should be to three significant figures – 44 400 cm^3.

2. No units are given in the stem of the question therefore you need to give units with your answer. You could therefore give your answer with units as 44 400 cm^3 or as 44·4 litres.

Calculations that are taught as part of the course

Most calculations come in Chemistry in Society, but they can be applied in contexts that may relate to Nature's Chemistry. There is one very simple calculation that is covered in Chemical Changes and Structure – the calculation of relative rate for a chemical reaction.

Relative rate of a reaction

At National 5 you will have dealt with average rate; at Higher you go on to look at relative rate. When we look at rate it is usually expressed as the rate at which a product is formed or a reactant is used up. So for a gas being produced it might be expressed as $cm^3 s^{-1}$ (cubic centimetres per second) or for an acid being used up it might be $mol\, l^{-1}\, s^{-1}$, that is, how the concentration is changing. The important aspect is that rate is always inversely proportional to time.

$$Rate \propto \frac{1}{time}$$

When we think about it, it is fairly obvious – the longer the time taken, the slower the rate of a chemical reaction and vice versa. The reciprocal of time is known as the relative rate and has units, s^{-1}.

$$Relative\ rate\ =\ \frac{1}{time} \qquad or \qquad Time\ for\ reaction\ =\ \frac{1}{relative\ rate}$$

Questions about relative rate often involve looking at a graph of relative rate or time for reactions against some variable such as temperature or concentration and deducing either the relative rate or the time for reaction.

Q Hydrogen peroxide can react with potassium iodide to produce water and iodine.

A student carried out an experiment to investigate the effect of changing the concentration of potassium iodide on reaction rate. The results are shown below.

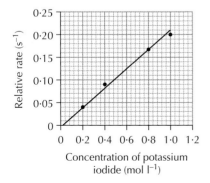

Calculate the time taken, in s, for the reaction when the concentration of potassium iodide used was $0.6\ mol\, l^{-1}$.

A To answer this question we simply read the relative rate from the graph.

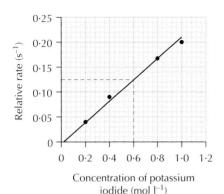

> **Top tip**
>
> Make sure you use your ruler to draw a line from the x-axis to the graph and then from the graph to the y-axis to ensure you obtain an accurate value for the relative rate.

Relative rate = $0 \cdot 125 \text{ s}^{-1}$

Time for reaction $= \dfrac{1}{\text{relative rate}}$

$= \dfrac{1}{0 \cdot 125}$

$= 8 \text{ s}$

Calculations using balanced equations

Balanced equations show the mole ratios of reactants and products. They can therefore be used to calculate the quantity of reactant(s) needed to produce an amount of product or the quantity of product produced from a given amount of reactants. Many questions test your ability to use balanced equations in this way.

Q Copper reacts with hot concentrated sulfuric acid to produce sulfur dioxide:

$Cu(s) + H_2SO_4(\ell) \longrightarrow CuSO_4(aq) + SO_2(g) + 2H_2O(\ell)$

Calculate the volume, in litres, of sulfur dioxide gas that would be produced when 10 grams of copper are reacted with excess concentrated sulfuric acid.

(Take the volume of 1 mole of sulfur dioxide gas to be 24 litres.)

A From the balanced equation we can see that one mole of copper gives one mole of sulfur dioxide gas. To calculate the volume we must first calculate the moles of copper reacted. This will produce the same number of moles of sulfur dioxide gas. It is simply then a case of multiplying this number by the molar volume (24) to give the final answer.

$$Cu(s) + H_2SO_4(\ell) \longrightarrow CuSO_4(aq) + SO_2(g) + 2H_2O(\ell)$$

1 mole   1 mole

10 g

$\dfrac{10}{63 \cdot 5}$ mol $\rightarrow$ $\dfrac{10}{63 \cdot 5}$ mol (moles of SO_2)

$$= \left(\dfrac{10}{63 \cdot 5} \right) \times 24 \text{ litres}$$

$$= 3 \cdot 8 \text{ (litres)}$$

Limiting reactants and reactants in excess

When we carry out a chemical reaction, very rarely will the reactants be in exact mole ratio amounts. There will normally be more of one particular reactant. Some of this reactant will remain when the reaction finishes.

A very simple example: if we want to prepare some copper sulfate crystals, we can do it by adding copper carbonate powder to dilute sulfuric acid. We add the copper carbonate until no more reacts. We know this is the case when no more carbon dioxide gas is given off and some of the insoluble copper carbonate remains unreacted in the bottom of the beaker.

The copper carbonate is said to be **in excess** – we added more than was needed to react with all the acid.

The volume of carbon dioxide gas given off in the reaction would have been determined by how much acid was used up rather than by the mass of copper carbonate added to the beaker.

The dilute acid would be described as **the limiting reactant**. That is, the reactant that determines how much product would have been formed.

Questions involving this type of calculation will either ask you to calculate the reactant in excess or the limiting reactant.

Q A student prepared a sample of methyl cinnamate from cinnamic acid and methanol.

cinnamic acid	+	methanol	$\rightarrow$	methyl cinnamate	+	water
mass of one mole		mass of one mole		mass of one mole		
= 148 g		= 32 g		= 162 g		

6·5 g of cinnamic acid was reacted with 2·0 g of methanol.

Show, by calculation, that cinnamic acid is the limiting reactant. (One mole of cinnamic acid reacts with one mole of methanol.)

A This question is simply a case of working out the quantities of reactants in moles. There will be fewer moles of the limiting reactant.

	cinnamic acid	methanol
Mass	6·5 g	2·0 g
Moles	$\dfrac{6·5}{148}$	$\dfrac{2·0}{32}$
	$= 0·044$ mol	$= 0·0625$ mol

We are told that one mole of cinnamic acid reacts with one mole of methanol. There are more moles of methanol so methanol is in excess. Cinnamic acid is therefore the limiting reactant.

Atom economy and percentage yield

It is important in the chemical industry to know how much product a reaction sequence is likely to produce. Both atom economy and percentage yield are indicators of the efficiency of reactions and reaction sequences.

Atom economy indicates the potential amount of the desired product that can be obtained by a particular synthetic route. A reaction with a low atom economy will mean that there are other products of the reaction that will either need to be disposed of or sold on as bi-products.

You will know from your studies in Higher Chemistry that not all reactions go to completion. Percentage yield is a measure of how the actual yield compares to the theoretical yield for a process.

Remember: the formulae used to calculate atom economy and percentage yield are given in the Data Booklet.

$$\% \text{ yield} = \frac{\text{Actual yield}}{\text{Theoretical yield}} \times 100$$

$$\% \text{ atom economy} = \frac{\text{Mass of desired product(s)}}{\text{Total mass of reactants}} \times 100$$

Q Methanamide, $HCONH_2$, is widely used in industry to make nitrogen compounds. It is also used as a solvent as it can dissolve ionic compounds.

In industry, methanamide is produced by the reaction of an ester with ammonia.

$HCOOCH_3$	$+$	NH_3	$\rightarrow$	$HCONH_2$	$+$	CH_3OH
mass of one mole		mass of one mole		mass of one mole		mass of one mole
$=60 \cdot 0$ g		$=17 \cdot 0$ g		$=45 \cdot 0$ g		$=32 \cdot 0$ g

Calculate the atom economy for the production of methanamide.

A This is simply a case of plugging the numbers into the formula given in the Data Booklet.

The mass of desired product is 45 g. The total mass of reactants is $60 + 17 = 77$ g.

The % atom economy is therefore

$$\frac{45}{77} \times 100 = 58 \cdot 4\%$$

The question went on to ask candidates to calculate a percentage yield.

Q In the lab, methanamide can be prepared by the reaction of methanoic acid with ammonia.

$HCOOH$	$+$	NH_3	$\rightleftharpoons$	$HCONH_2$	$+$	H_2O
mass of one mole		mass of one mole		mass of one mole		mass of one mole
$=46 \cdot 0$ g		$=17 \cdot 0$ g		$=45 \cdot 0$ g		$=18 \cdot 0$ g

When 1·38 g of methanoic acid was reacted with excess ammonia, 0·945 g of methanamide was produced.

Calculate the percentage yield of methanamide.

Show your working clearly.

A The advice to '**Show your working clearly**' is given in order that, if you don't end up with the correct answer, partial marks can be awarded for correct working or if you show understanding of a concept.

The first thing you have to do is **work out the theoretical yield**. This is simple proportion.

According to the equation 46·0 g of methanoic acid should give 45·0 g of methanamide.

$$HCOOH \rightarrow HCONH_2$$

$$1 \text{ mol} \rightarrow 1 \text{ mol}$$

$$46.0 \text{ g} \rightarrow 45.0 \text{ g}$$

$$\text{By proportion } 1.38 \text{ g} \rightarrow 45.0 \times \frac{1.38}{46.0} \text{ g}$$

$$= 1.35 \text{ g (Theoretical yield)}$$

The actual yield is given as 0·945 g.

The % yield is therefore $\left(\dfrac{0.945}{1.35} \right) \times 100 = 70\%$

Check that the answer you end up with is reasonable – remember a percentage yield has to be less than 100%.

Enthalpy change calculations

As well as knowing the amount of product that is likely to be obtained during a reaction, it is important to know about the energy changes that take place during a reaction. This is referred to as the **enthalpy change for the reaction.**

Experimentally, the enthalpy change can be measured using a simple calorimeter system. In the lab, you may have used a system similar to the one in the question below to measure the enthalpy of combustion of methanol.

Enthalpy of combustion
This is one enthalpy calculation that you may be asked to carry out that depends on knowing the definition. Remember from your definitions that the enthalpy of combustion is the energy released when **one mole** of a substance burns.

Q Butan-1-ol, C_4H_9OH, can be blended with petrol. The enthalpy of combustion of butan-1-ol can be measured experimentally. An apparatus that can be used to carry out the experiment is shown.

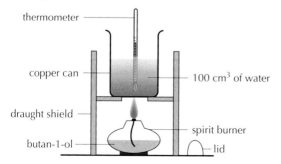

In the experiment, burning 0·64g of butan-1-ol raised the temperature of the water in the can from 20 °C to 41 °C.

Use these results to calculate an experimental value for the enthalpy of combustion, in kJ mol^{-1}, of butan-1-ol.

A Again the equation you need to use to work out the energy produced by the burning butan-1-ol is given in the Data Booklet:

$$E_h = cm\Delta T$$

c is the specific heat capacity of water and is given in the Data Booklet as 4·18 kJ kg^{-1} °C^{-1}. What does this mean?

If we have 1 kilogram of water and want to raise its temperature we need to put in 4·18 kilojoules of energy for every degree Celsius of temperature rise.

We can use this equation to calculate the heat energy transferred to the water.

m = 0·1 kg (100 cm^3 of water has a mass of 100 g = 0·1 kg. A very common mistake by candidates is to use the change in mass of the substance being burned here. It's the mass of the liquid being heated that is needed.)

$$\Delta T = 21 \text{ °C}$$

Using the equation we can calculate the energy transferred to the water:

E_h = cm ΔT

= 4·18 kJ kg^{-1} °C^{-1} × 0·1 kg x 21 °C

= 8·8 kJ

> Notice how the units cancel out, leaving kilojoules.

The question tells us that this energy was produced when 0·64 g of butan-1-ol burned.

1 mole of butan-1-ol has a mass of 74 g.

We now just have to use proportion to work out the energy that would be produced when 1 mole of butan-1-ol is burned.

0·64 g → 8·8 kJ

74 g → $8·8 \times \dfrac{74}{0·64}$

= 1018 kJ (An answer of 1015 kJ is obtained if you carry and use your answer from your calculator, i.e. use 8·778 kJ.)

Notice that the question asks you to **calculate the enthalpy of combustion.**

We really need to give our answer properly by including the negative sign to show that the process is exothermic, and giving the units as kilojoules per mole:

$$\Delta H_{comb} \text{ butan-1-ol} = -1018 \text{ kJ mol}^{-1}$$

Next we have a question that looks at a practical use of a chemical reaction producing heat energy.

Q Self-heating cans may be used to warm drinks such as coffee.

When the button on the can is pushed, a seal is broken, allowing water and calcium oxide to mix and react.

The reaction produces solid calcium hydroxide and releases heat.

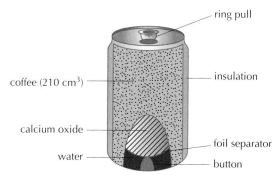

The equation for this reaction is:

$$CaO(s) \quad + \quad H_2O(\ell) \quad \rightarrow \quad Ca(OH)_2(s) \qquad \Delta H = -65 \text{ kJ mol}^{-1}$$

Calculate the mass, in grams, of calcium oxide required to raise the temperature of 210 cm³ of coffee from 20 °C to 70 °C.

Show your working clearly.

A The question might appear difficult but remember the examiners are asking you to show your working clearly. You can obviously get marks for showing some understanding.

What do you have? You've got a volume of water that you can change to a mass (remember 1 litre of water has a mass of 1 kilogram) and a temperature change:

$$m = 0.21 \text{ kg} \quad \Delta T = 50 \text{ °C}$$

Now you have to use the equation.

$$E_h = cm \, \Delta T$$
$$= 4.18 \text{ kJ kg}^{-1} \text{ °C}^{-1} \times 0.21 \text{ kg} \times 50 \text{ °C}$$
$$= 43.89 \text{ kJ}$$

We are told: 1 mole of CaO(s) reacting with water gives out 65 kJ of energy. 1 mole of CaO has a mass of 56·1 g.

So now it's proportion. We only need 43·89 kJ of energy so we know our answer will be less than 56·1 g:

$$56·1 \text{ g} \quad \rightarrow \quad 65 \text{ kJ}$$

$$56·1 \times \frac{43·89}{65} \quad \rightarrow \quad 43·89 \text{ kJ}$$

$$= \underline{37·9 \text{ g}}$$

Bond enthalpy calculations

Bond enthalpy values can also be used to estimate the enthalpy change for chemical reactions that take place in the gas phase. In the gas phase any intermolecular forces have been broken and will not therefore need to be considered.

The Data Booklet lists tables of bond enthalpy values and mean bond enthalpy values.

What's the difference?

Bond Enthalpies		Mean Bond Enthalpies	
Bond	Enthalpy / kJ mol^{-1}	Bond	Mean Enthalpy / kJ mol^{-1}
H – H	436	Si – Si	226
O = O	498	C – C	348
N ≡ N	945	C = C	612
F – F	159	C ≡ C	838
Cl – Cl	243	C – C (aromatic)	518
Br – Br	194	H – O	463
I – I	151	H – N	388
H – F	570	C – H	412
H – Cl	432	C – O	360
H – Br	366	C = O	743
H – I	298	C – F	484
		C – Cl	338
		C – Br	276
		C – I	238

A bond enthalpy value is the energy needed to break a bond in a diatomic molecule. A mean bond enthalpy value is given where there can be different atoms and groups attached to the atoms given in the table. For instance, the energy needed to break an oxygen-to-hydrogen bond will be slightly different if the OH is attached to another hydrogen, such as if it is in water or if it is attached to an ethyl group in ethanol. The mean bond an enthalpy is just an average value for the energy to break the bond or given out when the bond forms.

Again the key to getting this type of question right is setting it down properly.

Q Methanol can be converted to methanal as shown.

$$H-\overset{\displaystyle H}{\underset{\displaystyle H}{C}}-O-H \quad \longrightarrow \quad \overset{\displaystyle H}{\underset{\displaystyle H}{\diagdown}} C=O \quad + \quad H-H$$

Using bond enthalpy and mean bond enthalpy values from the Data Booklet, calculate the enthalpy change, in kJ mol^{-1}, for the reaction.

A When we examine the structures equation we can identify the bonds that are broken and the new bonds that are formed during the reaction. Three bonds need to be broken and two new bonds form. Energy is required to break bonds – the process is endothermic; energy is given out when new bonds form – this is exothermic.

Bonds broken	Energy required (kJ)	Bonds formed	Energy given out (kJ)
C — H × 1	412	C = O × 1	−743
C — O × 1	360	H — H × 1	−436
O — H × 1	463		
Total energy required	1235	Total energy given out	−1179

Enthalpy from the reaction $= 1235 + (-1179)$ kJ mol^{-1}

$$= +56 \text{ kJ mol}^{-1}$$

You may not want to draw a table but try to set your working out in a way that lets the marker check what you are doing. Partial marks were given in this question for evidence of having used the correct bond enthalpy values. If you didn't end up with the correct answer, you would need to have written them down in order to be given credit.

Hess's Law calculations

The enthalpy change for some reactions can't be determined experimentally.

Take for example the following reaction.

$$C(s) \quad + \quad 2H_2(g) \quad \rightarrow \quad CH_4(g)$$

Carbon and hydrogen won't react together to form methane, therefore the enthalpy change, ΔH, can't be determined experimentally.

Hess's Law allows us to use other reactions to work out what the enthalpy change might be.

Hess's Law states that the overall energy change for a reaction will be the same no matter the chemical pathway taken.

We know that carbon and hydrogen will burn to give carbon dioxide and water and that methane burns to give carbon dioxide and water as well. One mole of carbon and two moles of hydrogen will burn to give the same products as one mole of methane.

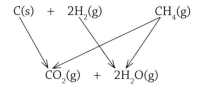

The enthalpy change for these reactions can be found experimentally.

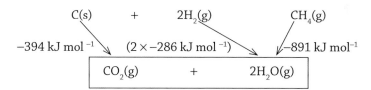

The enthalpy change then for carbon and hydrogen reacting to give methane would be the same as going from carbon and hydrogen to carbon dioxide and water and then from carbon dioxide and water to methane. The enthalpy change in going from carbon dioxide and water to methane would just be the reverse of the energy in going from carbon dioxide and water to methane.

By Hess's Law, ΔH for $\qquad$ C(s) $\quad + \quad 2H_2(g) \quad \rightarrow \quad CH_4(g) \quad$ would be

$$= -394 + (-572) + 891$$

$$= -75 \text{ kJ mol}^{-1}$$

The calculation can be done by manipulating the combustion equations for carbon, hydrogen and methane:

C(s) $\quad + O_2(g) \quad \rightarrow \quad CO_2(g)$ $\qquad\qquad \Delta H = -394 \text{ kJ mol}^{-1}$

$H_2(g) \ + \frac{1}{2}O_2(g) \ \rightarrow \ H_2O(\ell)$ $\qquad\qquad \Delta H = -286 \text{ kJ mol}^{-1}$

$CH_4(g) + 2O_2(g) \quad \rightarrow \quad CO_2(g) + 2H_2O(\ell) \qquad \Delta H = -891 \text{ kJ mol}^{-1}$

The target equation is:

$$C(s) \quad + \quad 2H_2(g) \quad \rightarrow \quad CH_4(g)$$

The first thing we have to do is reverse the combustion equation for methane to get methane molecules on the correct side. Because we reverse the equation we have to change the sign of the enthalpy change.

$CO_2(g) + 2H_2O(\ell) \rightarrow CH_4(g) + 2O_2(g) \qquad \Delta H = +891 \text{ kJ mol}^{-1}$

Now we need to get rid of the $CO_2(g)$ on the reactant side by adding in the combustion of carbon equation. The carbon dioxides will cancel out, and one of the oxygens will as well:

$\cancel{CO_2}(g) + 2H_2O(\ell) \rightarrow CH_4(g) + \cancel{2}O_2(g) \qquad \Delta H = +891 \text{ kJ mol}^{-1}$

$C(s) \quad + \cancel{O_2}(g) \quad \rightarrow \cancel{CO_2}(g) \qquad\qquad \Delta H = -394 \text{ kJ mol}^{-1}$

We now need to cancel out the waters on the reactant side by adding the hydrogen combustion equation. This will also cancel out the remaining oxygen. This time we need to double the equation so we have to multiply the enthalpy change by 2.

$\cancel{CO_2}(g) + \cancel{2H_2O}(\ell) \rightarrow CH_4(g) + \cancel{2O_2}(g) \qquad \Delta H = +891 \text{ kJ mol}^{-1}$

$C(s) \quad + \cancel{O_2}(g) \quad \rightarrow \cancel{CO_2}(g) \qquad\qquad \Delta H = -394 \text{ kJ mol}^{-1}$

$2H_2(g) + \cancel{O_2}(g) \quad \rightarrow \cancel{2H_2O}(\ell) \qquad\qquad \Delta H = -572 \text{ kJ mol}^{-1}$

Now when we add the three equations we get the target equation. Adding the different enthalpy changes gives us the enthalpy change for the reaction:

$$C(s) + 2H_2(g) \rightarrow CH_4(g) \ \Delta H = -75 \text{ kJ mol}^{-1}$$

Diagrammatic forms of Hess's Law might be used in the multiple choice section of the paper, whereas equations are likely to be used in Section 2 of the paper.

Q Calcium hydroxide solution can be formed by adding calcium metal to excess water.

Solid calcium hydroxide would form if the exact molar ratio of calcium to water is used. The equation for the reaction is:

$$Ca(s) + 2H_2O(\ell) \rightarrow Ca(OH)_2(s) + H_2(g)$$

Calculate the enthalpy change, in kJ mol^{-1}, for the reaction above by using the data shown below.

$H_2(g) + \frac{1}{2}O_2(g) \rightarrow H_2O(\ell)$ $\qquad\qquad\qquad$ $\Delta H = -286$ kJ mol^{-1}

$Ca(s) + O_2(g) + H_2(g) \rightarrow Ca(OH)_2(s)$ $\qquad\qquad$ $\Delta H = -986$ kJ mol^{-1}

A In this question you are only given two equations to manipulate to achieve the target equation.

Target equation: $\qquad$ $Ca(s) + 2H_2O(\ell) \rightarrow Ca(OH)_2(s) + H_2(g)$

Examining the two equations you have been given, the second one has $Ca(OH)_2(s)$ written on the product side so is in the correct direction.

The first equation however has $H_2O(\ell)$ written on the product side. We will therefore need to reverse this equation. We will also need to multiply by a factor of 2 since the target equation has 2 moles of water. This means we will need to multiply the enthalpy change by 2 and change its sign.

$Ca(s) + O_2(g) + H_2(g) \rightarrow Ca(OH)_2(s)$ $\qquad\qquad$ $\Delta H = -986$ kJ mol^{-1}

Reversing the first equation and multiplying by 2:

$2H_2O(\ell) \rightarrow O_2(g) + 2H_2(g)$ $\qquad\qquad\qquad$ $\Delta H = +572$ kJ mol^{-1}

Adding the two equations together gives the target equation. Adding the enthalpy changes gives the overall enthalpy change.

$Ca(s) + 2H_2O(\ell) \rightarrow Ca(OH)_2(s) + H_2(g)$ $\qquad$ $\Delta H = \underline{-414 \text{ kJ mol}^{-1}}$

Titrations

If you have done National 5 Chemistry you will have covered titration calculations in the section, Acids and Bases. At Higher, titration calculations are extended to include redox titrations. Of all the calculations that you will carry out at Higher these are probably the most demanding. It is therefore vital that you follow a systematic approach.

We will start by looking at an example where you are given the full redox equation for a reaction.

Q 25·0 cm^3 samples of the diluted soft drink were titrated with Fehling's solution which had a Cu^{2+} concentration of 0·0250 mol l^{-1}.

The average volume of Fehling's solution used in the titrations was 19·8 cm^3.

$$C_6H_{12}O_6 \quad + \quad 2Cu^{2+} \quad + \quad H_2O \quad \rightarrow \quad C_6H_{12}O_7 \quad + \quad 2Cu^+ \quad + \quad 2H^+$$

| reducing sugar | Fehling's solution | | | | | |

Calculate the concentration, in mol l^{-1}, of reducing sugars present in the diluted sample of the soft drink.

A We will use two different methods to carry out the calculation.

Method 1
Using the relationship: Number of moles = concentration × volume in litres

The question tells us that the volume of Fehling's solution used was 19·8 cm^3 and that the Cu^{2+}(aq) was 0·025 mol l^{-1}.

Step 1
Work out the moles of Cu^{2+}(aq) used.

$v = 19·8$ cm$^3 = 0·0198$ l

$c = 0·025$ mol l^{-1}

$\qquad$ Number of moles $= 0·0198 \times 0·025$

$\qquad\qquad\qquad = 0·000495$ mol or $4·95 \times 10^{-4}$ mol

Step 2
Use the mole ratios from the redox equation to work out the moles of reducing sugar.

The equation shows that 1 mol reducing sugar reacts with 2 mol Cu^{2+}(aq)

There is half as much reducing sugars.

Moles of reducing sugars $= 0·0002475$ mol or $2·475 \times 10^{-4}$ mol

Step 3

Use $n = c \times v$ to work out the concentration

$n = 0.0002475$ mol

$v = 25$ cm^3 = 0.025 l

$$c = \frac{0.0002475}{0.025}$$

$= 0.0099$ mol l^{-1} or 9.9×10^{-3} mol l^{-1}

Method 2
Using a formula
Using this method we don't have to change cm^3 to litres as long as we use the same volume units on each side of the equation. This is the same formula that worked for titrations involving acids and bases.

$$\frac{\left[(conc \times vol)\; reactant\, A\right]}{\left[Balancing\; number\; reactant\, A\right]} = \frac{\left[(conc \times vol)\, reactant\, B\right]}{\left[Balancing\; number\; reactant\, B\right]}$$

The balancing numbers are the mole ratios for each reactant from the redox equation.

A good way to start is to make a table of what you know from the question or scribble the information on your paper below each reactant in the redox equation.

	Reducing sugars	$Cu^{2+(aq)}$
Conc (mol l^{-1})	?	0.025
Vol (cm^3)	25	19.8
Bal no.	1	2

Then it is a case of plugging the numbers into the formula.

$$\frac{\left[(conc \times 25)\, reducing\; sugars\right]}{1} = \frac{\left[(0.025 \times 19.8)\, copper\, ions\right]}{2}$$

$$Conc = \frac{(0.025 \times 19.8)}{(25 \times 2)}$$

$$= 0.0099 \text{ mol l}^{-1}$$

Both methods obviously give the same answer. Some candidates prefer method 1, others prefer method 2. It is up to you to choose the method you feel most comfortable with.

Now let's look at an example with a slight twist.

Q The concentration of sodium hypochlorite in swimming pool water can be determined by redox titration.

Step 1

A 100·0 cm^3 sample from the swimming pool is first reacted with an excess of acidified potassium iodide solution forming iodine.

$$NaOCl(aq) + 2I^-(aq) + 2H^+(aq) \rightarrow I_2(aq) + NaCl(aq) + H_2O(\ell)$$

Step 2

The iodine formed in step 1 is titrated using a standard solution of sodium thiosulfate, concentration 0·00100 mol l^{-1}. A small volume of starch solution is added towards the end point.

$$I_2(aq) + 2Na_2S_2O_3(aq) \rightarrow 2NaI(aq) + Na_2S_4O_6(aq)$$

Calculate the concentration, in mol l^{-1}, of sodium hypochlorite in the swimming pool water, if an average volume of 12·4 cm^3 of sodium thiosulfate was required.

A You have been given a lot of information and the question may seem fairly daunting. What we have been asked for, the concentration of hypochlorite, can't be found directly from the redox equation. This is where a systematic approach setting out your working clearly really helps.

We will use the first method we described above, that is, use the relationship:

Number of moles = concentration × volume in litres

The first thing we can do is calculate the moles of thiosulfate used in the titration.

We are given the volume and concentration of thiosulfate used in the titration so can use the relationship n = c × v to work out the number of moles of thiosulfate used.

$c = 0·00100$ mol l^{-1}

$v = 12·4$ cm$^3 = 0·0124$l

n (no. of moles of thiosulfate) $= 0·00100 \times 0·0124$

$= 0·0000124$ mol

This can be left as 0·0000124 mol or changed to scientific notation 1·24 x 10^{-5}

Now that we know the number of moles of thiosulfate we can work out the number of moles of iodine it reacted with.

Looking at the titration equation, the second equation you are given, we can see that 1 mole of iodine reacts with 2 moles of thiosulfate. Therefore there must be half as much iodine.

So, moles of iodine $= 0.0000062$ or 6.2×10^{-6} mol

If we look back at the first equation we can see that the 1 mole of hypochlorite reacts to give 1 mole of iodine.

The moles of iodine found by titration must be the same as the moles of hypochlorite in 100 cm^3 of water from the swimming pool.

For the hypochlorite we now have:

$n = 6.2 \times 10^{-6}$ mol

$v = 100$ cm$^3 = 0.1$ l

Now we use $n = c \times v$ to work out the concentration. Re-arranging the equation we get $c = \dfrac{n}{v}$

Concentration $= \dfrac{6.2 \times 10^{-6}}{0.1} = \underline{6.2 \times 10^{-5} \text{ mol l}^{-1}}$

Using the formula approach

The formula approach allows us to calculate the concentration of iodine from the redox equation. It is the same as carrying out steps 1 and 2 by Method 1.

	Iodine	thiosulfate
Conc (mol l^{-1})	?	0.00100
Vol (cm^3)	100	12.4
Bal no.	1	2

$$\frac{\left[(conc \times 100)iodine\right]}{1} = \frac{\left[(0.001 \times 12.4)thiosulfate\right]}{2}$$

$$Conc = \frac{(0.001 \times 12.4)}{(100 \times 2)}$$

$$= 0.000062 \text{ or } 6.2 \times 10^{-5} \text{ mol l}^{-1}$$

Notice you don't have to give the units as they are given in the question.

To calculate the concentration of hypochlorite we would still need to carry out the same final calculation as shown for the first method.

If we look back at the first equation we can see that the 1 mole of hypochlorite reacts to give 1 mole of iodine.

The moles of iodine found by titration must be the same as the moles of hypochlorite in 100 cm^3 of water from the swimming pool.

For the hypochlorite we now have:

$n = 6 \cdot 2 \times 10^{-6} \text{ mol}$

$v = 100 \text{ cm}^3 = 0 \cdot 1 \text{ l}$

Now we use $n = c \times v$ to work out the concentration. Re-arranging the equation we get

$$c = \frac{n}{v}$$

$$\text{Concentration} = \frac{6 \cdot 2 \times 10^{-6}}{0 \cdot 1} = \underline{6 \cdot 2 \times 10^{-5} \text{ mol l}^{-1}}$$

When doing titration calculations, choose the method that you are most comfortable with, but remember **whichever method you use, show your working clearly**.

Picking up points in problem solving

CHAPTER 6

This chapter covers:

- It pays to be methodical
- When you are given rules – follow them!
- Scribble on your paper to help you work things out
- Never assume you know the answer

Be prepared for the unfamiliar

Do you like puzzles? Can you follow instructions? Every year there are a number of questions that are set in very unfamiliar chemistry contexts. Many of the questions associated with these contexts simply ask you to apply your knowledge but some test your ability to process information, make generalisations and predict what would happen in similar situations. For many of these questions you wouldn't need to have studied chemistry to get the answer. These are questions that anyone can attempt but it obviously is of benefit to have studied chemistry.

These are questions that can appear off-putting because of the unfamiliarity of the situations or the volume of information you appear to be being given. All the information you need to be able to answer the question will be right there in front of you. It will simply be a case of you processing the information by following instructions or making sense of information that you are given and applying this information in a new situation. The more familiar you are with this type of question from past papers, the less likely you are to be put off when you face them in your exam.

Remember when tackling these questions, the first thing is to **keep calm** – if the chemistry is unfamiliar to you, it's likely to be unfamiliar to everyone else as well.

It pays to be methodical

Always read the whole question through. That means all the information you are given and what you are being asked to do. With some questions it will be fairly obvious but for some others there will be a bit of puzzling to be done.

Let's look at some examples.

This first question asks you to complete a flow chart from information.

Start by reading the passage through. The industrial process will likely be unfamiliar but it's just about different chemicals. You need to process the information and enter it correctly into the flow chart.

Q Sodium carbonate is used in the manufacture of soaps, glass and paper as well as the treatment of water.

One industrial process used to make sodium carbonate is the Solvay process.

The Solvay process involves several different chemical reactions.

It starts with heating calcium carbonate to produce carbon dioxide, which is transferred to a reactor where it reacts with ammonia and brine. The products of the reactor are solid sodium hydrogencarbonate and ammonium chloride which are passed into a separator.

The sodium hydrogencarbonate is heated to decompose it into the product sodium carbonate along with carbon dioxide and water. To recover ammonia the ammonium chloride from the reactor is reacted with calcium oxide produced by heating the calcium carbonate. Calcium chloride is a by-product of the ammonia recovery process.

Using the information above, complete the flow chart by adding the names of the chemicals involved.

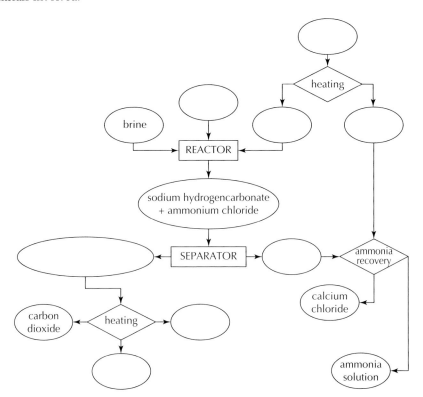

A Start by linking sections of the information to specific parts of the diagram.

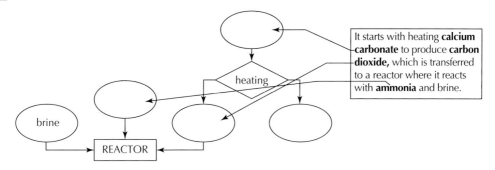

We can link three of the top four ovals to chemicals in the first sentence. We don't have the fourth oval. If you know your chemistry you can work out that this must be calcium oxide. But you don't need to worry about it as there is information later about recovering ammonia that allows us to fill it in.

Now use the fragment, 'the products of the reactor are solid sodium hydrogencarbonate and ammonium chloride which are passed into a separator' to help you fill in two more ovals.

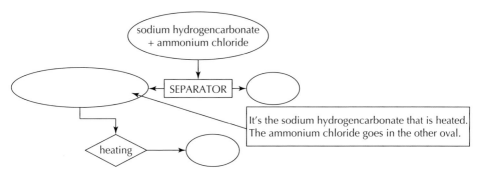

The products of heating are sodium carbonate, carbon dioxide and water. Since carbon dioxide is given, the other two ovals must be for sodium carbonate and water.

There is one last oval to fill in. It's the one we said not to worry about at the beginning.

'To recover ammonia the ammonium chloride from the reactor is reacted with calcium oxide produced by heating calcium carbonate.'

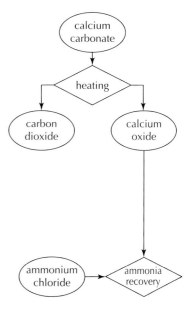

Answer

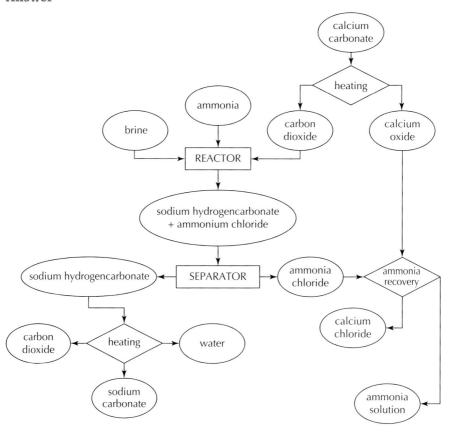

Here are some more examples where you need to be methodical, but specifically methodical with your puzzling.

Q Methanol reacts with compound X, in an addition reaction, to form methyl tertiary-butyl ether, an additive for petrol.

$$CH_3OH(g) + X \longrightarrow H_3C-\underset{\underset{CH_3}{|}}{\overset{\overset{OCH_3}{|}}{C}}-CH_3$$

methyl tertiary-butyl ether

Suggest a structure for compound X.

A In this question methanol is being added to a compound X. We can see where the CH_3O from the methanol has gone and we can deduce that the hydrogen must have gone on to one of the other carbons. We need to take these away from the structure. This leaves:

$$H-\underset{\underset{H}{|}}{\overset{\overset{|}{}}{C}}-\underset{\underset{CH_3}{|}}{\overset{\overset{|}{}}{C}}-CH_3$$

Now to get the final answer we need to connect the two carbons with a double bond. There's a clue that you need a double bond in your structure – we are told it's an **addition** reaction.

So our final answer should be:

$$H_2C=\underset{\underset{CH_3}{|}}{C}-CH_3$$

Q Partial hydrolysis of another pentapeptide molecule gave a mixture of three smaller peptide molecules with the following amino acid sequences:

leucine-glycine-valine
isoleucine-leucine
glycine-valine-serine

Write the amino acid sequence for the original pentapeptide molecule.

A This is quite a simple one to work out really if you are a problem solver.

Write the first fragment:

leucine-glycine-valine

Now slide the next two fragments until amino acid fragments with the same name are aligned:

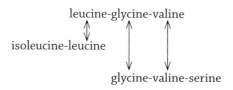

leucine-glycine-valine

isoleucine-leucine

glycine-valine-serine

Now it's just a case of writing the sequence down starting at the left.

Answer: isoleucine-leucine-glycine-valine-serine

Q Geraniol is one of the compounds found in perfume. It has the following structural formula and systematic name.

H—C—C=C—C—C—C=C—C—H

3,7-dimethylocta-2,6-dien-1-ol

Linalool can also be present. Its structural formula is shown.

H—C—C=C—C—C—C—C=C—H

State the systematic name for linalool.

A In this question there's nothing unfamiliar about the structure. You can identify double bonds, methyl side chains and hydroxyl groups. You just have to make sense of the name.

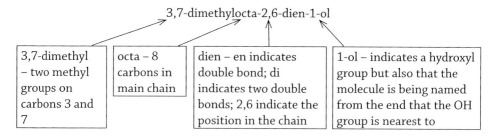

3,7-dimethylocta-2,6-dien-1-ol

3,7-dimethyl – two methyl groups on carbons 3 and 7	octa – 8 carbons in main chain	dien – en indicates double bond; di indicates two double bonds; 2,6 indicate the position in the chain	1-ol – indicates a hydroxyl group but also that the molecule is being named from the end that the OH group is nearest to

Now we need to apply this to the target molecule.

The molecule still has 8 Cs in the main chain, therefore the octa will remain.

OH is on the third carbon from the right and sixth counting from the left, so we therefore count from the right. The name will end in '-3-ol'.

Two carbon-to-carbon double bonds starting on carbons 1 and 6 in the chain. The name will include '-1,6-dien-' (remember to put in the di bit).

Two methyl groups on carbons 3 and 7 therefore '3,7-dimethyl'.

So, putting it all together gives the answer:

3,7-dimethylocta-1,6-dien-3-ol.

Q The table below shows the duration of numbness for common anaesthetics.

Name of anaesthetic	Structure	Duration of numbness / minutes
procaine		7
lidocaine		96
mepivacaine		114
anaesthetic **X**		

Estimate the duration of numbness, in minutes, for anaesthetic **X**.

A **It's easy to see something that looks very complex and panic but remember – don't panic!**

Some teachers refer to these types of questions as 'the Friday frighteners' – the type of questions they give their class at the end of a period or the end of the week as a bit of a change from their normal work. Teachers are trying to train you not to be put off when you face something that looks difficult. The information is there – just be methodical.

In this question, look at the information you are given about lidocaine and mepivacaine. The left-hand side of each structure is the same. It's the right-hand side of the molecules that are different. The right-hand side of lidocaine has a branched structure and that of mepivacaine has a ring structure. The duration of numbness is 18 minutes more for mepivacaine.

Now look at procaine and anaesthetic X – the left-hand side of the structure is the same; the right-hand side changes from the branched structure to the ring structure, just like lidocaine and mepivacaine. It would be reasonable to assume that the numbness might increase by the same amount (18 minutes) and therefore go from 7 minutes to 25 minutes.

But that's not the only way of looking at it. We can think in terms of factors rather than straight differences. Going from 96 to 114 is an increase of $\frac{114}{96}$ as a factor. On your calculator this comes out as 1·1875. It is reasonable to suggest that 7 minutes might be increased by this same factor, which would give a time of 8·3 minutes.

Both 25 minutes and 8·3 minutes are acceptable answers to give to this question.

Q 5-butyl-4-methyltetrahydrofuran-2-ol is a flavour compound found in whisky stored in oak barrels.

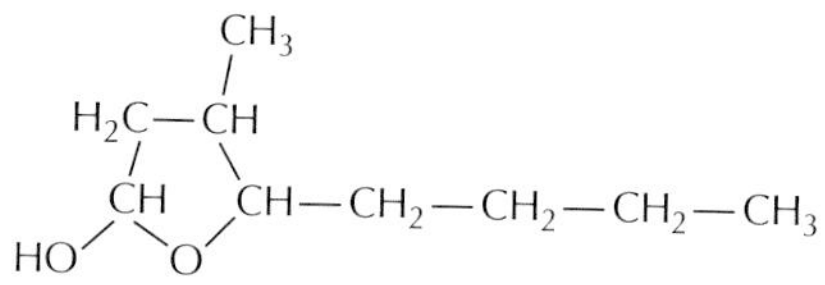

$$\begin{array}{c} CH_3 \\ | \\ H_2C-CH \\ / \qquad \backslash \\ CH \qquad CH-CH_2-CH_2-CH_2-CH_3 \\ HO \qquad O \end{array}$$

5-butyl-4-methyltetrahydrofuran-2-ol

Write the systematic name for the compound shown below.

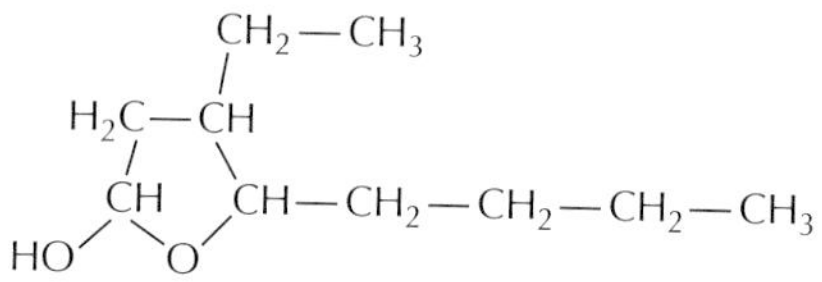

$$\begin{array}{c} CH_2-CH_3 \\ | \\ H_2C-CH \\ / \qquad \backslash \\ CH \qquad CH-CH_2-CH_2-CH_2-CH_3 \\ HO \qquad O \end{array}$$

A What a complicated name! You aren't given a rule to follow for this one.

You might start by looking at the first structure and trying to link parts of the name to the structure. This will certainly help.

We can work out from the name that the ring is numbered from the oxygen. The carbon with the hydroxyl group attached is position 2 in the ring; the carbon with the methyl group attached is position 4; and the carbon with the butyl group attached is position 5 in the ring.

5-butyl-4-methyltetrahydrofuran-2-ol

However, is this the smartest thing to do? Perhaps not!

Read the whole question and look at all the information.

Look at the molecules. The only difference between the example and the molecule you are being asked to name is that the one you are being asked to name has an ethyl group attached to the ring instead of a methyl group.

The only bits of the molecules that are different. The butyl group and the –ol part are unaffected.

5-butyl-4-**methyl**tetrahydrofuran-2-ol

The only part of the name that will be affected will be that it will have an ethyl group rather than a methyl group.

The name is therefore:

5-butyl-4-ethyltetrahydrofuran-2-ol

When you are given rules – follow them!

A number of problem-solving questions will give rules and an example. By relating the rules to the example given you can then apply the rules in a related situation.

Q In solution, amino acid molecules can form zwitterions when a hydrogen ion moves from the carboxyl group onto the amino group.

For example,

$$
\begin{array}{cc}
\text{H} & \text{O} \\
| & \| \\
\text{H}_2\text{N}-\text{C}-\text{C}-\text{OH} \\
| \\
\text{H}
\end{array}
\longrightarrow
\begin{array}{cc}
\text{H} & \text{O} \\
| & \| \\
\text{H}_3\overset{+}{\text{N}}-\text{C}-\text{C}-\text{O}^- \\
| \\
\text{H}
\end{array}
$$

glycine   glycine zwitterion

Draw the zwitterion produced by the amino acid serine.

$$
\begin{array}{c}
\text{O} \quad \text{H} \quad\quad \text{H} \\
\| \quad\ | \quad\quad\ | \\
\text{HO}-\text{C}-\text{C}---\text{C}-\text{OH} \\
\quad\quad\ | \quad\quad\ | \\
\quad\quad \text{NH}_2 \quad \text{H}
\end{array}
\longrightarrow
$$

serine   serine zwitterion

A You've probably never heard of a zwitterion but it's just a case of following the rule. This example also shows **you need to read carefully**. The mistake that lies in wait for the unwary candidate who quickly looks at the example but doesn't read the rule carefully is to remove the hydrogen from the oxygen on the right of the molecule. This is the wrong oxygen, however, and you should first read the instruction – the H^+ has to come from the carboxyl group. Now identify the groups involved in both the example and then in the question.

$$
\begin{array}{c}
\quad\quad\quad\quad\quad \text{O} \quad \text{H} \quad\quad \text{H} \\
\quad\quad\quad\quad\quad \| \quad\ | \quad\quad\ | \\
\text{HO}-\text{C}-\text{C}---\text{C}---\text{OH} \\
\quad\quad\quad\quad\ | \quad\quad\ | \\
\quad\quad\quad\quad \text{NH}_2 \quad \text{H}
\end{array}
$$

This is the carboxyl group.

This is the amino group.

Now it's just a case of removing an H^+ from the carboxyl, leaving a negative charge, and adding an H^+ to the amino group.

$$
\begin{array}{c}
\quad\quad\quad \text{O} \quad \text{H} \quad\quad \text{H} \\
\quad\quad\quad \| \quad\ | \quad\quad\ | \\
{}^-\text{O}-\text{C}---\text{C}---\text{C}---\text{OH} \\
\quad\quad\quad\quad\ | \quad\quad\ | \\
\quad\quad\quad \ {}^+\text{NH}_3 \quad \text{H}
\end{array}
$$

The H^+ has moved from the carboxyl group to the amino group.

Here are more examples where it is just a case of following the rules.

Q Borane (BH_3) is used to synthesise alcohols from alkenes.

The reaction occurs in two stages.

Stage 1 Addition Reaction
The boron atom bonds to the carbon atom of the double bond which already has the most hydrogens **directly** attached to it.

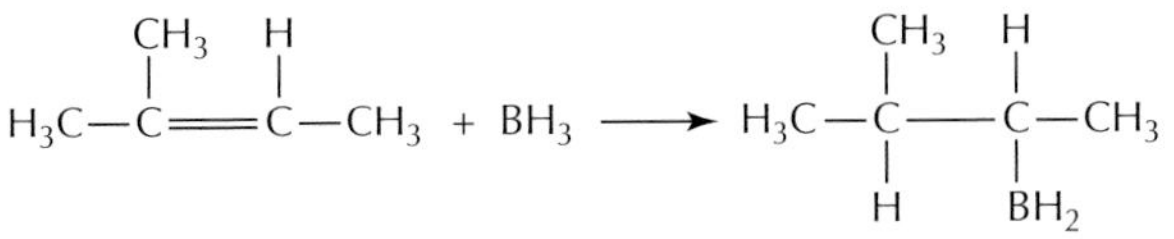

$$H_3C-\overset{\overset{\displaystyle CH_3}{|}}{C}=\overset{\overset{\displaystyle H}{|}}{C}-CH_3 + BH_3 \longrightarrow H_3C-\overset{\overset{\displaystyle CH_3}{|}}{\underset{\underset{\displaystyle H}{|}}{C}}-\overset{\overset{\displaystyle H}{|}}{\underset{\underset{\displaystyle BH_2}{|}}{C}}-CH_3$$

Stage 2 Oxidation Reaction
The organoborane compound is oxidised to form the alcohol.

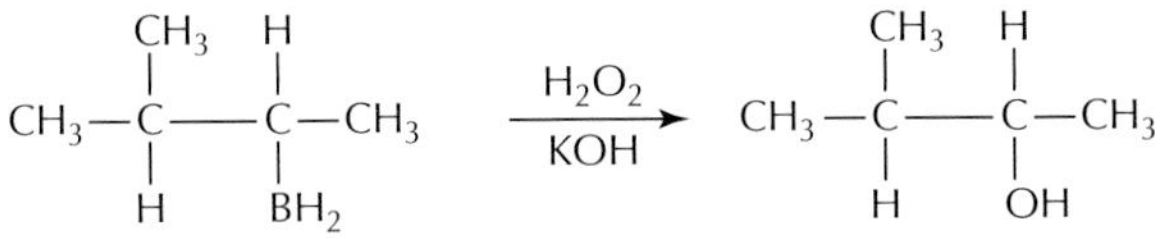

$$CH_3-\overset{\overset{\displaystyle CH_3}{|}}{\underset{\underset{\displaystyle H}{|}}{C}}-\overset{\overset{\displaystyle H}{|}}{\underset{\underset{\displaystyle BH_2}{|}}{C}}-CH_3 \xrightarrow[\text{KOH}]{H_2O_2} CH_3-\overset{\overset{\displaystyle CH_3}{|}}{\underset{\underset{\displaystyle H}{|}}{C}}-\overset{\overset{\displaystyle H}{|}}{\underset{\underset{\displaystyle OH}{|}}{C}}-CH_3$$

$$CH_3-CH_2-CH_2-\overset{\overset{\displaystyle CH_3}{|}}{C}=\overset{\overset{\displaystyle H}{|}}{C}-H$$

Draw a structural formula for the **alcohol** which would be formed from the alkene shown on the right.

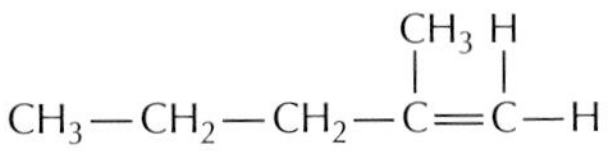

A

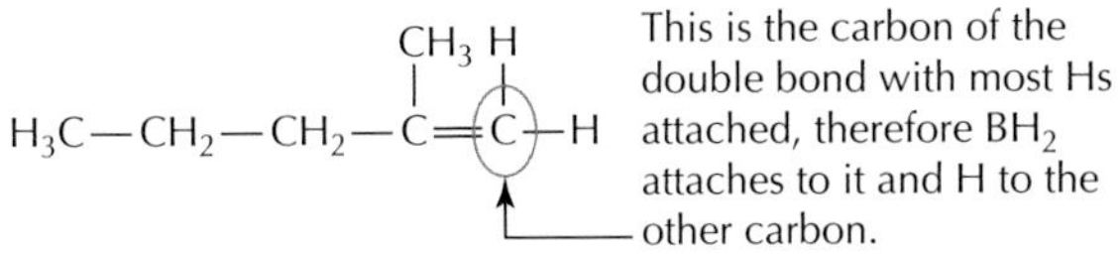

$$H_3C-CH_2-CH_2-\overset{\overset{\displaystyle CH_3}{|}}{C}=\overset{\overset{\displaystyle H}{|}}{C}-H$$

This is the carbon of the double bond with most Hs attached, therefore BH_2 attaches to it and H to the other carbon.

The intermediate is therefore:

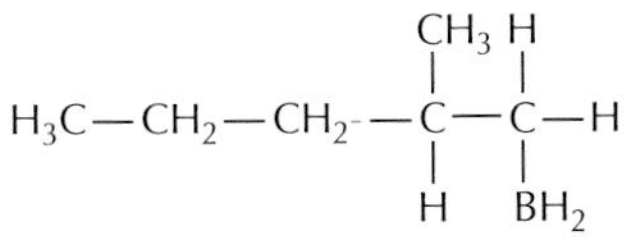

$$H_3C-CH_2-CH_2-\overset{\overset{\displaystyle CH_3}{|}}{\underset{\underset{\displaystyle H}{|}}{C}}-\overset{\overset{\displaystyle H}{|}}{\underset{\underset{\displaystyle BH_2}{|}}{C}}-H$$

Now replace the BH_2 with OH to get the alcohol. Simple, isn't it!

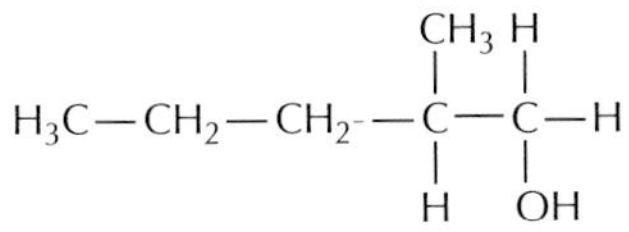

$$H_3C-CH_2-CH_2-\overset{\overset{\displaystyle CH_3}{|}}{\underset{\underset{\displaystyle H}{|}}{C}}-\overset{\overset{\displaystyle H}{|}}{\underset{\underset{\displaystyle OH}{|}}{C}}-H$$

Q The products formed when an explosive substance decomposes can be predicted by applying the Kistiakowsky-Wilson rules. These rules use the number of oxygen atoms in the molecular formula to predict the products.

In the example below these rules are applied to the decomposition of the explosive RDX, $C_3H_6N_6O_6$

Rule number	Rule	Atoms available in $C_3H_6N_6O_6$	Apply rule to show products
1	Using oxygen atoms from the formula convert any carbon atoms in the formula to carbon monoxide.	$3 \times C$	3CO formed
2	If any oxygen atoms remain, convert H atoms in the formula to water.	$3 \times O$ remain	$3H_2O$ formed
3	If any oxygen atoms still remain then convert CO formed to CO_2.	No more oxygen left	No CO_2 formed
4	Convert any nitrogen atoms in the formula to N_2.	$6 \times N$	$3N_2$ formed

Decomposition equation:

$$C_3H_6N_6O_6(s) \rightarrow 3CO(g) + 3H_2O(g) + 3N_2(g)$$

By applying the same set of rules, complete the equation for the decomposition of the explosive PETN, $C_5H_8N_4O_{12}$.

$C_5H_8N_4O_{12}(s) \rightarrow$

A Here's another one to follow the rules.

Let's apply the rules to $C_5H_8N_4O_{12}$

Rule 1: Convert 5 Cs to 5 COs   5 of the 12 Os used, 7 Os left.

Rule 2: 8 Hs use up 4Os making 4 H_2O 4 of the 7 Os used, 3 Os left.

Rule 3: 3 Os remain, 3 of the 5 COs can be changed to 3 CO_2; 2 COs remain unchanged.

Rule 4: 4 N atoms form 2 N_2 molecules.

At the end you have $4H_2O$; $3CO_2$; 2CO and $2N_2$

Answer: $C_5H_8N_4O_{12}(s) \rightarrow 3CO_2(g) + 2CO(g) + 4H_2O(g) + 2N_2(g)$

As a final check you can count that you have the same number of atoms on each side. If they are different you have made a mistake!

Scribble on your paper to help you work things out

This next question asks you to follow a rule but it will also help if you scribble a couple of lines on your paper.

Q In solution, sugar molecules exist in an equilibrium in straight-chain and ring forms.

To change from the straight-chain form to the ring form, the oxygen of the hydroxyl on carbon number 5 joins to the carbonyl carbon. This is shown below for glucose.

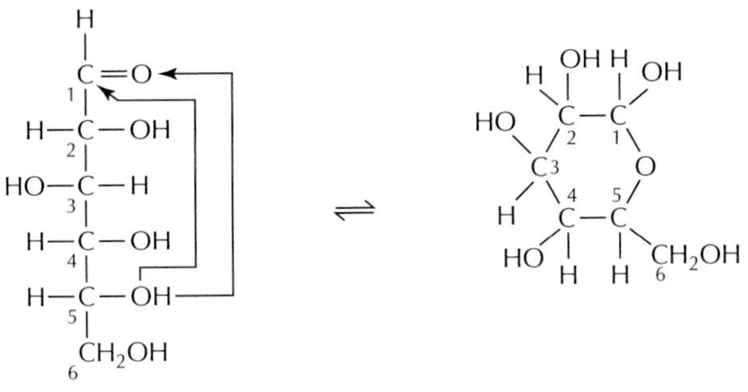

glucose

Draw the structure of a ring form for fructose.

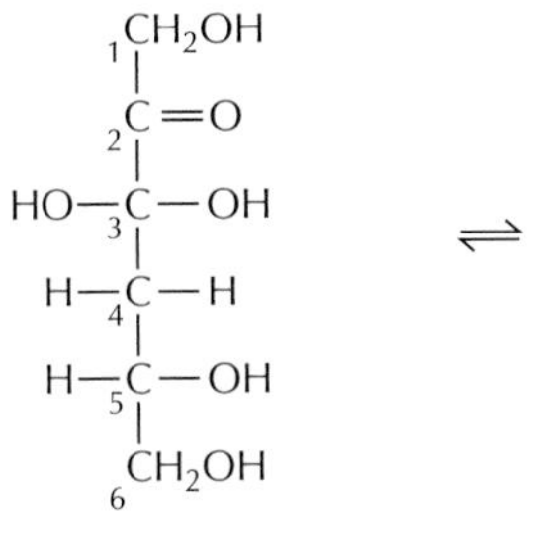

fructose

A Again, follow the rule (but scribble the links showing where the bonds form on the initial structure you are given). Scribbling on the structure you are given helps you first work out that the structure will end up with a ring of 5 atoms (4 carbons and 1 oxygen atom). Now you can look at each atom in the ring and work out what is attached. Carbon 2 will have a CH_2OH and an OH attached. Carbon 5 will have a CH_2OH and an H attached. Carbons 3 and 4 have an H and an OH attached.

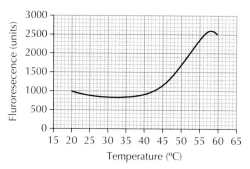

Q Another protein in egg white is conalbumin. The temperature of a conalbumin / dye mixture is gradually increased. The fluorescence is measured and a graph is produced.

The melting temperature is the temperature at which the fluorescence is halfway between the highest and lowest fluorescence values.

Determine the melting temperature, in °C, for this protein.

A Now you have to be prepared to draw on your paper to help you get the right answer. If you are given a graph, draw lines to work out your answer. And, by the way, use a ruler! It's very easy to make a mistake by going up or down a box when you are doing this without one.

Answer: The highest fluorescence is 2600 units and the lowest just under 900 units (860). Halfway between would be approximately 1730–1750 units.

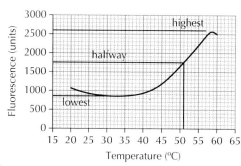

Using a ruler to help you find the reading accurately gives a temperature of between 50 and 51 °C as the melting temperature.

And now for one last thing...

Never assume you know the answer

Q Electronegativity values can be used to predict the type of bonding present in substances.

The type of bonding between two elements can be predicted using the diagram below.

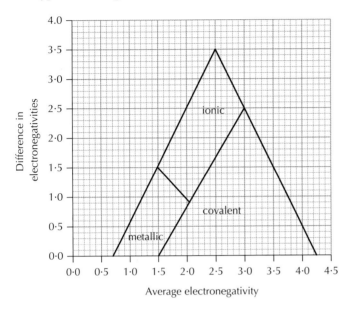

The diagram can be used to predict the bonding in tin iodide.

Electronegativity of tin $= 1 \cdot 8$

Electronegativity of iodine $= 2 \cdot 6$

Average electronegativity $= 2 \cdot 2$

Difference in electronegativity $= 0 \cdot 8$

Predict the type of bonding in tin iodide.

A You might be tempted to think: 'Oh this is easy! Tin is a metal therefore tin iodide will be ionic'. Would examiners really ask you to predict this if that was the case?

You might think you know the answer but chemistry isn't as clear cut as that, and you need to follow the rules. Now you need to be a bit more precise than scribbling on your paper. You need to draw lines to see where they cross on the diagram.

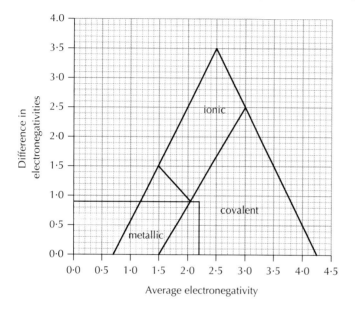

When you draw the lines for electronegativity difference and average electronegativity on the diagram you can see that tin chloride is predicted to be covalent.

So, in summary, if you want to pick up marks in problem solving:

- Don't panic because it is unfamiliar
- Be methodical
- Follow the rules
- Be prepared to scribble on your paper to work things out
- Never assume you know the answer.

Tackling open questions

This chapter covers:

- Can you prepare for open questions?
- Highlight important words

One feature of the new chemistry exam papers is open questions. These are questions where there is no single correct answer. Rather, the questions examine your understanding of chemistry relating to a particular situation. The questions have an allocation of 3 marks and your answer will be marked according to whether the answer displays good, reasonable, limited or no understanding. Your answer doesn't need to be perfect and it is quite possible for your answer to be completely different from that of another candidate but for both answers to be awarded 3 marks.

The following marking criteria are used to mark open questions.

If an answer is thought to be:

Good	3 marks
Reasonable	2 marks
Limited	1 mark
No Understanding	0 marks

A good answer doesn't need to cover every aspect that you can think of, or make three points. Rather, the statements you make need to be relevant and correct.

Can you prepare for open questions?

The answer is, 'Yes, you can'!

Answering open questions is often about recognising which areas might be relevant to the question and making links between the different parts of chemistry you have covered. It is worth noting that the chemistry that you describe should be chemistry you have covered in the Higher course rather than chemistry that may have been covered at lower levels.

The way you study can help you develop skills that will aid you in answering open questions. Don't just read through your notes. It is good sometimes to stop and test yourself by writing down what you know.

For instance you can start with a blank sheet of paper. First you write down a topic in the middle of the page and then create a diagram – we'll use the term 'memory map', although other terms are used (network diagram, spider diagram, mind map) – to put down on paper what you have learned from the Chemistry course.

Let's take an example before we look at a couple of open questions.

Alcohols form a significant part of the Higher Chemistry course. After revising this area of work, you want to see how much you remember and begin to create a memory map. You write the word alcohols in the middle of a blank sheet and the first thing you remember is that they contain the hydroxyl functional group and that they can be divided into primary, secondary and tertiary alcohols. So you draw this with some key information on the sheet below the box in which you wrote the word 'alcohols'.

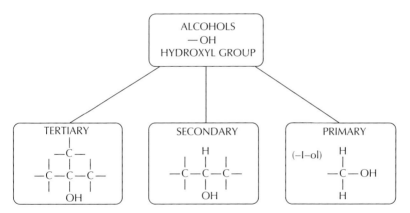

You then start to build up the diagram with things you know about alcohols.

Very quickly you should be able to link in things you know about other aspects of alcohol chemistry until you have a diagram that links lots of the alcohol chemistry you have covered.

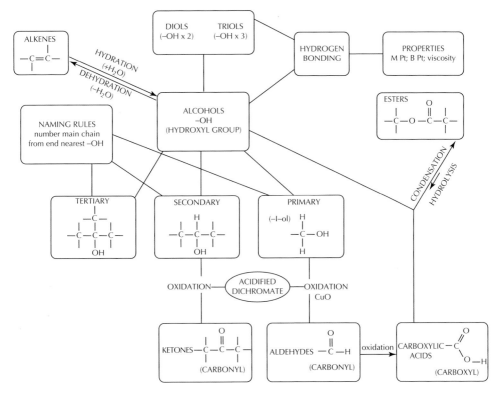

Don't throw away any diagrams you create. You can pin them up somewhere to remind you or to refer to from time to time.

This is the type of activity that will help you make links between different aspects of chemistry. This tip is mentioned in the **10 top tips for revision** that are given in Chapter 11.

Highlight important words

Now let's look at a couple of open questions. Remember, there is no single correct answer.

Q Patterns in the periodic table

The periodic table is an arrangement of all the known elements in order of increasing atomic number. The reason why the elements are arranged as they are in the periodic table is to fit them all, with their widely diverse physical and chemical properties, into a logical pattern.

Periodicity is the name given to regularly occurring similarities in physical and chemical properties of the elements.

Some Groups exhibit striking similarity between their elements, such as Group 1, and in other Groups, the elements are less similar to each other, such as Group 4, but each Group has a common set of characteristics.

Adapted from *Royal Society of Chemistry, Visual Elements* (rsc.org)

Using your knowledge of chemistry, comment on similarities and differences in the patterns of physical and chemical properties of elements in both Group 1 and Group 4.

A We are given a lot of clues about how we should answer this question. Simply highlighting key phrases in the introductory text and asking yourself fairly obvious questions will help you answer the question.

Let's consider some phrases that you might highlight.

Periodicity is the name given to regularly occurring similarities in physical and chemical properties of the elements.

Some Groups exhibit striking similarity between their elements, such as Group 1, and in other Groups the elements are less similar to each other, such as Group 4, but each Group has a common set of characteristics.

Now let's think about thoughts you might have and questions you might ask yourself.

- Periodicity – What periodic properties have I learned about at Higher?
  - Answer: covalent radius; ionisation energy; electronegativity. Phew, I know about periodicity! (You might even have done a network map of periodicity, and if you have you certainly will have these three properties written down.)
- Group 1 – Can I explain trends in periodic properties for groups?
- Group 4 – Less similar? What's different?
- Common characteristics – What does every group have in common?

There's a lot to write about!

Firstly answer that last point. What does every group have in common?

- Elements in the same group have the same outer shell configuration – Group 1 has 1 electron in the outer shell; Group 4 has 4 electrons.

And now write about each of the groups mentioned.

- Group 1 – Alkali metals (discounting hydrogen), all very reactive metals; each element has the largest atom of its period, all have lowest first ionisation energy due to atom size and lowest electronegativity of their period. Atom size increases down the group; first ionisation energy and electronegativity decrease down the group.
- Group 4 – What's different? (What do you have that tells you? Not your knowledge ...it's your Data Booklet!) Immediately you see that the group is made up of non-metals (C and Si) and metals (Ge, Sn, Pb). Are the trends in the periodic properties the same? Lead's first ionisation energy doesn't follow the pattern – it's higher than tin's. Electronegativity doesn't follow the pattern either. Silicon's is lower than germanium's.

You can of course go into defining first ionisation energy and electronegativity and why Alkali metal atoms are the largest of the period – as long as you are showing understanding at Higher level.

Now ask yourself, should you have been able to give a limited, reasonable or good answer? Or would you not have had a clue? The fact that you are reading this book indicates that at the very least your answer should be reasonable, and should gain at least 2 of the 3 marks available.

Let's look at another example. Let's highlight key phrases to focus on.

Q The chemical industry creates an immense variety of products which impact on virtually every aspect of our lives. Industrial scientists, including chemical engineers, production chemists and environmental chemists, carry out different roles to maximise the efficiency of industrial processes.

Using your knowledge of chemistry, comment on what industrial scientists can do to maximise profit from industrial processes and minimise impact on the environment.

A This question focuses straight in on the key area 'getting the most from reactants'. When you covered this area in class you will have thought about:

- Factors influencing process design. These include:
  - o availability, sustainability and cost of feedstock(s)
  - o opportunities for recycling
  - o energy requirements
  - o marketability of by-products
  - o product yield.

- Environmental considerations such as:
  - minimising waste
  - avoiding the use or production of toxic substances
  - designing products that will biodegrade if appropriate.

There is lots to write about relating to the bullet points above. If you have done any sort of network chart you'll surely have a link from maximising efficiency to atom economy and percentage yield.

You can make the points such as:

- Ideally you want a process that has high atom economy and high percentage yield.
- A process with a low atom economy means you have other products. It's okay if these by-products are useful for other purposes and can be sold on but if they have to be disposed of then this can create problems.
- A high atom economy process may have a low percentage yield, therefore the chemical plant must be designed so that unused reactants can be recycled.

You can talk about reaction conditions such as rate of reaction versus percentage yield; controlling temperature and pressure affecting percentage yield. For some reactions a low temperature might be capable of producing a high percentage yield but the rate of reaction may be too slow. Chemists need to monitor reaction conditions to ensure that optimum conditions are maintained.

If you can give examples, all the better.

For environmental impact you can talk about biodegradability. You can also talk about systems that scrub acidic gases from emissions.

To summarise, the best advice regarding answering open questions is:

- to consider creating your own memory maps when revising key areas in order that you can better understand how bits of the chemistry course link together
- to focus in on the key phrases when answering the questions. Remember these might come in the introductory passage or be incorporated into the question and there is nothing wrong with you using a highlighter on your paper in the exam to help you do this
- to use diagrams and equations where appropriate to illustrate your answer
- remember there is no single correct answer.

Researching Chemistry

This chapter covers:

- Familiarity with apparatus
- Familiarity with techniques
- Processing experimental results

These are the three key aspects developed during this part of the course. Approximately 10 marks in your written final exam will test your knowledge of the skills you develop in the Researching Chemistry unit of the course.

The Higher Chemistry Course Support Notes, which can be downloaded from the SQA website, detail the apparatus and techniques you should be familiar with and the skills needed to process experimental results that you should develop during the course.

Familiarity with apparatus

You should be familiar with the following pieces of apparatus:

- conical flask – this is used when carrying out titrations and other general laboratory work
- beaker – the volume markings on a beaker only give a rough indication of volume
- measuring cylinder – sufficiently accurate for preparatory work such as making up solution, but not sufficiently accurate for analytical work
- delivery tubes – when drawing these in diagrams, they need to be open at the end
- dropper
- test tubes / boiling tubes – test tubes are narrower than boiling tubes and should not be heated directly using a Bunsen flame as the liquid being heated will tend to spurt. Boiling tubes are sufficiently broad to allow a liquid to be carefully heated using a Bunsen flame
- evaporating basin

- pipette with safety filler – used to measure / deliver a specified volume of liquid accurately
- burette – used to determine the volume of liquid used accurately
- volumetric flask – a flask with a narrow neck and a graduation mark that allows a specific volume of solution to be prepared
- funnel
- thermometer.

Being familiar with the above means you would be expected to recognise, draw and understand the limitations of use of the apparatus.

Researching Chemistry questions can appear to be quite simple but don't be lulled into a sense of complacency. This next question proved to be extraordinarily difficult for candidates.

In SQA 2016 New Higher Q11 (a) (i) candidates were asked to: 'Draw a diagram of a volumetric flask'.

The question shows how much we, as teachers, often take for granted. Sometimes we assume that because students have used a volumetric (standard) flask they will remember what it looks like and be able to draw it.

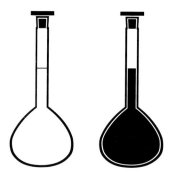

What candidates were required to do was draw a flask with a long narrow neck that had a single graduation mark on the neck of the flask. This question was extremely poorly answered.

Candidates at Higher should appreciate why a volumetric flask has this shape. It is shaped like this to improve the accuracy when making a standard solution. The flask has a narrow neck in order that the liquid in the flask will give a nicely curved meniscus. When the flask is filled so that the bottom of the meniscus touches the mark, the flask will contain exactly the amount specified. A wider neck would mean that the meniscus would not be curved, making it more difficult to fill the flask accurately.

Most often your knowledge of apparatus and its use can be examined by asking you to complete a diagram. You can also be asked how to use or set up pieces of apparatus.

These next two questions examine your knowledge of how to use equipment properly.

Q Some students carried out an investigation of fruit drinks to determine their vitamin C content. The following steps were followed in each experiment:

Step 1 A 20·0 cm^3 sample of fruit drink was transferred to a conical flask by pipette.

Step 2 A burette was filled with a standard iodine solution.

Step 3 The fruit drink sample was titrated with the iodine.

Step 4 Titrations were repeated until concordant results were obtained.

The burette, pipette and conical flask were all rinsed before they were used.

Tick the appropriate boxes below to show which solution should be used to rinse each piece of glassware.

Glassware used	Rinse with water	Rinse with iodine	Rinse with fruit drink
pipette			
burette			
conical flask			

A This question tested candidates' understanding of how to prepare apparatus for use. Apparatus has to be clean and also should not affect the accuracy of a procedure.

In titrations the conical flask needs to be clean, and therefore needs to be rinsed with water before use. Burettes and pipettes should also be rinsed with water to ensure they are clean. However, if the burette or pipette is wet then the water will affect the concentration of the liquid being used. Burettes and pipettes must therefore be rinsed with the liquid they will contain.

The expected correct response would be:

Glassware used	Rinse with water	Rinse with iodine	Rinse with fruit drink
pipette			✓
burette		✓	
conical flask	✓		

Q Describe fully a method that the student could have used to accurately measure the mass of 10·0 cm^3 of each sucrose solution.

A There are two aspects to this question – the accurate measurement of volume and the accurate measurement of mass. Many candidates didn't appear to see the words 'the mass' and described how to accurately measure 10·0 cm^3 of each sucrose solution. **Another case of when it pays to read the question carefully!**

The answer that was needed was that a clean, dry flask should be weighed empty and then 10·0 cm^3 should be transferred to the flask using a 10·0 cm^3 pipette. The flask should then be reweighed. The weight of the sucrose solution would be the difference between the two weighings.

Top tip

It sometimes helps to try to visualise what you do when carrying out a procedure. Try to imagine carrying out the instruction.

Familiarity with techniques

The techniques you should be familiar with are discussed below.

Filtration
You are only expected to know about simple filtration systems.

Use of a balance / simple gravimetric analysis
For example, weighing a precipitate or weighing by difference. Gravimetric analysis is a technique that allows the mass of a substance or a concentration of a substance to be determined by precipitating the substance and weighing the compound formed. Weighing by difference is a method of reducing the errors associated with weighing. A few grams of

the substance to be weighed out are placed into a weighing bottle, the mass recorded, and the approximate amount required is tapped out into a small beaker.

The weighing bottle is picked up with a loop of paper or tongs to prevent transferring oils from our hands to the weighing bottle. The weighing bottle with the remaining substance is reweighed. The mass transferred is given by the difference in the two recorded masses.

Safe methods of heating
Bunsen burners, water baths or heating mantles – when are particular methods used?

- Bunsen burners are used to heat non-flammable liquids.
- Water bath arrangements are used when the reaction takes place at temperatures below 100 °C and the liquids being used are flammable or need to be heated for a prolonged period of time.
- Heating mantles are used when prolonged heating at temperatures above 100 °C are required.

Volumetric analysis
Volumetric analysis is the use of a standard solution of accurately known concentration to find the concentration of another solution or of a substance in a solution, for example vitamin C in orange juice. Volumetric analysis requires understanding of how to prepare a standard solution and of how to carry out a titration.

Preparation of a standard solution
An accurately weighed mass of solute is transferred into a beaker. If the mass has been weighed on a watch glass, the surface of the clock glass should be rinsed into the beaker. The solid is then dissolved in a small volume of water. The liquid is transferred to a volumetric flask and the beaker rinsed and the rinsings added to the flask. The solution is then made up to the mark with distilled / deionised water. The last few drops are added from a dropping pipette until the bottom of the meniscus touches the mark. The flask is then stoppered and inverted several times to thoroughly mix the solution.

Titration
A burette is filled with one of the solutions to be titrated. Known volumes of the other solution are transferred to a conical flask using a pipette. An indicator is added (unless the reaction is self-indicating). A rough titre can be carried out to give an indication of the end point. The solution from the burette is added until the end point is reached. The solution is added slowly as the end point is neared. The end point is indicated by a colour change in the reaction mixture.

Methods for following rates of reactions
Rates of reaction questions are most often linked to collision theory and the calculation of relative rates. The rate of a chemical reaction can be followed in several ways: measuring

the rate at which a gas is given off from a reaction or measuring how the mass of the reaction mixture changes as the reaction proceeds.

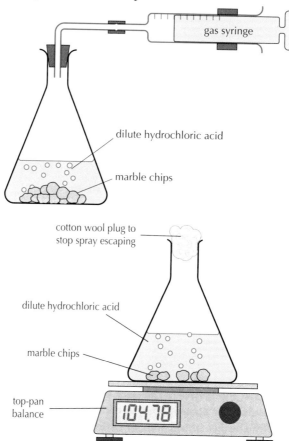

Chromatography

You are not required to know specific details regarding different chromatographic methods. However you would be expected to explain results of chromatography experiments.

Q On a chromatogram, the retention factor R_f for a substance can be a useful method of identifying the substance.

$$R_f = \frac{\text{distance moved by the substance}}{\text{maximum distance by the solvent}}$$

The structure of the pentapeptide methionine enkephalin was investigated.

A sample of the pentapeptide was completely hydrolysed into its constituent amino acids and this amino acid mixture was applied to a piece of chromatography paper and placed in a solvent.

The chromatogram obtained is shown on the right.

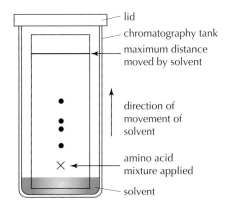

Suggest why only four spots were obtained on the chromatogram of the hydrolysed pentapeptide.

A You don't need to know anything about this particular type of chromatography. You can use your knowledge that separation depends on size of molecules and polarity. The initial part of the question explained that a pentapeptide was made up of five amino acid units. If two of the amino acids have similar sizes and polarity they are likely to travel similar distances and their spots will overlap. The question could also be answered in terms of the information given in the question by stating that two of the amino acids had the same R_f value.

Organic analysis of structure

These are the tests you would use to determine particular structural features of organic compounds. In particular you are expected to be able to distinguish between primary, secondary and tertiary alcohols and between aldehydes and ketones.

Primary and secondary alcohols can be oxidised to aldehydes and ketones respectively using hot copper(II) oxide or acidified dichromate.

Aldehydes can be oxidised to carboxylic acids using:

- Fehling's solution – colour change blue to orange-red.
- Acidified dichromate – colour change orange to green.
- Tollen's reagent – silver mirror formed.

Ketones cannot be oxidised further.

Degree of unsaturation

Any questions relating to degree of unsaturation are likely to be in the context of fats and oils. From National 5 you should know that bromine solution is decolourised by unsaturated compounds.

Q The following apparatus can be used to compare the degree of unsaturation of different oils.

Describe how this apparatus could be used to show that olive oil has a greater degree of unsaturation than coconut oil.

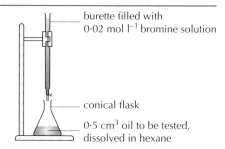

burette filled with 0·02 mol l^{-1} bromine solution

conical flask

0·5 cm^3 oil to be tested, dissolved in hexane

A As the bromine solution is added from the burette it will be decolourised as bromine molecules are added to the carbon-to-carbon double bonds of the oil. The bromine solution will continue to decolourise until all of the carbon-to-carbon double bonds have been used up. The end point will be indicated when the bromine solution is no longer decolourised. Since olive oil has a greater degree of unsaturation, a larger volume of bromine solution will need to be added to reach the end point.

Distillation

You would be expected to recognise a distillation set up. Note a Liebig condenser is not one of the listed pieces of apparatus. You would not be expected to draw this piece of apparatus.

Distillation

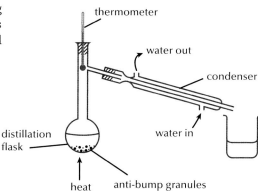

thermometer

water out

condenser

distillation flask

water in

heat   anti-bump granules

Solvent extraction

Similarly with solvent extraction, separating funnels are not listed as apparatus you need to know about.

Determining enthalpy

The diagram shows the apparatus you might use in the lab to determine the enthalpy of combustion of a fuel.

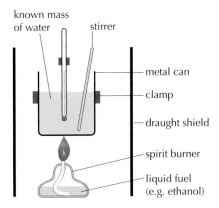

Processing experimental results

When you process experimental results you would be expected to:

Represent experimental data using a scatter graph and sketch lines or curves of best fit

Q The graph was plotted using the absorbance of different permanganate solutions.

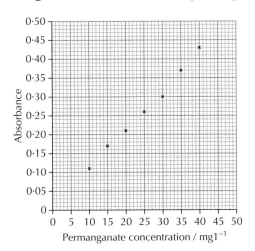

A sample of steel was reacted to give one litre of solution containing permanganate ions. The absorbance of the solution was 0·30.

Use your graph to determine the mass of manganese in the steel sample.

A In this question the candidates first had to determine the permanganate concentration for a solution with absorbance of 0·30.

Many candidates simply went up the y-axis to 0·30 and read across. They saw the plotted point and read down to the x-axis and then used a concentration of 30 mg l^{-1} in their calculation. **Candidates who did this missed one vital step!**

Candidates were expected to first of all **draw a best fit line**, then use this to determine the permanganate ion concentration.

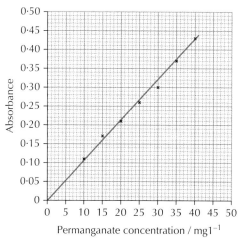

A best fit line gives a permanganate concentration of 28 mg l^{-1}. It's this figure that should have been used in the calculation.

Identify and eliminate rogue points from the analysis of results
You would also be expected to identify from a graph or a set of results any value that was inconsistent with the pattern or trend being shown by the other values. This type of value is described as a rogue value.

Q Standard ethanol solutions were used to produce a graph of density against concentration of ethanol, given as a percentage of alcohol by volume (% abv).

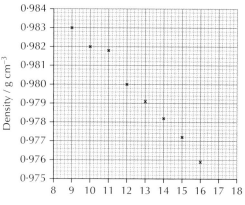

What is the concentration of ethanol, in units of % abv, in a solution of density 0.9818 g cm^{-3}?

A As with the last example a best fit line needs to be drawn. The plotted value for solution with density 0·9818 is obviously a rogue value. A best fit line would give a concentration of approximately 10·2 %.

Calculate averages (means) for experiments
This next question illustrates these last two points.

Q The mass of zinc in four 100 g samples taken from a cheese spread was measured.

Sample	Mass of Zn / mg
1	4·0
2	21·7
3	3·9
4	4·1

Calculate the average mass of Zn, in mg, in 100 g of this cheese spread.

A Looking at the table of results, sample 2 is quite obviously a rogue value since it is so different from the others.

The average mass of zinc would therefore be the average of the other three results:

Average mass of zinc = (4·0 + 3·9 + 4·1) / 3 = 4·0 mg

Another example of disregarding rogue results is disregarding a rough titre when calculating an average titre value. Titres are deemed to be concordant if the values are within 0·2 cm^3 of each other. Titre values outwith this tolerance can be deemed to be rogue values.

Show an appreciation of the relative accuracy of apparatus used to measure the volume of liquids
You are expected to know that the volume markings on beakers provide only a rough indication of volume and that while measuring cylinders generally provide sufficient accuracy for preparative work, for analytic work burettes, pipettes and volumetric flasks should be used.

Appreciate that when a measurement has been repeated, any variations in the value obtained give an indication of the reproducibility of the technique.

Know that the uncertainty associated with a measurement can be indicated in the form, *measurement ± uncertainty*. You would not, however, be expected to conduct any form of quantitative error analysis.

Every measurement has a degree of uncertainty.

When interpolating graphs you are normally allowed a half-box tolerance in any value that has to be read from a graph.

Carry out quantitative mole calculations
You need to be comfortable using the different mole relationships to carry out calculations.

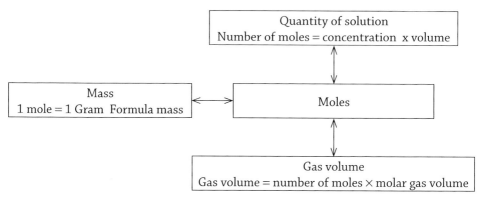

This next example links mass to gas volumes.

Q The explosive RDX, $C_3H_6N_6O_6$, is used in the controlled demolition of disused buildings.

During the reaction it decomposes as shown.

$$C_3H_6O_6N_6(s) \rightarrow 3CO(g) + 3H_2O(g) + 3N_2(g)$$

Calculate the volume, in litres, of gas released when 1·0 g of RDX decomposes.

Take the molar volume of the gases to be 24 litres mol^{-1}.

A Looking at the equation we can see that 1 mole of the explosive will produce 9 moles of gases, 3 of CO, 3 of H_2O, and 3 of N_2 (note that the state symbol for H_2O is (g) showing it is produced as a gas).

The first step is therefore to work out the number of moles of explosive used:

$$1 \text{ mole} = 1 \text{ Gram Formula mass}$$

The gram formula mass of RDX is 222 g.

Using $Number\ of\ moles = \dfrac{Mass}{Gram\ Formula\ Mass}$, $\dfrac{1}{222}$ moles of explosive were used.

Now we can work out moles of gas produced. Since 1 mole of explosive produces 9 moles of gas, $\dfrac{1}{222}$ moles would produce $\dfrac{9}{222}$ moles of gas.

We can now work out the volume.

$$Gas\ volume = number\ of\ moles \times molar\ gas\ volume$$

$$= \dfrac{9}{222} \times 24$$

$$= 0{\cdot}97\ litres$$

Being prepared for internal assessment

This chapter covers:

- Evidence gathering
- What can you be asked?
- So what should you know?
  - Chemical Changes and Structure
  - Nature's Chemistry
  - Chemistry in Society

Evidence gathering

Internal Assessment tests are likely to be an integral part of your course and can serve different purposes. They can be used by teachers to gauge your level of understanding and to help identify areas where you would benefit from additional practice. However, there are situations where centres may have to send off evidence of your abilities to the SQA.

Every year there are candidates who for reasons such as personal or family illness are either unable to sit the exam or are likely to underperform due to some external factor. In these circumstances centres can submit evidence to support the grade that they have predicted for you.

Evidence needs to be provided that you have good knowledge of the course content and that you have developed the problem-solving skills associated with the course.

You are expected to be able to display the following problem-solving skills. The ability to:

- make generalisation / predictions
- select information
- process information, including calculations, as appropriate
- analyse information.

Gathering the evidence

Centres can decide on the approach that will be taken in order to generate evidence that you have achieved these assessment standards.

Teachers are likely either to use tests covering a number of key areas of the course or a prelim exam that mirrors the content and structure of the final exam.

What can you be asked?

Questions in your final exam will be asked using **command words.** These will help you to know the type of response that will be required to gain the mark(s) for the question.

The command words that are being used include:

identify, **name**, **give**, or **state** – a single word or brief statement is all that is required to answer questions that begin with these command words.

> *Example: Name the type of reaction taking place when aldehydes change to carboxylic acids.*

The answer only requires the single word, **oxidation**.

describe – you must provide a brief statement.

> *Example: Describe a chemical test that will distinguish the aldehyde, propanal from the ketone, propanone.*

The answer would require you to describe how to carry out the test and the result you would expect:

A few drops of each compound should be added to about 5 cm³ of acidified potassium dichromate in separate test tubes and the test tubes placed in a hot water bath. The propanal will change the yellow dichromate solution to green. No reaction will take place using propanone.

Simply answering *'Use acidified potassium dichromate solution'* wouldn't be sufficient to gain you the mark.

explain – you must relate cause and effect; in other words you are giving an answer that says why something is as it is, or why something happens.

> *Example: The following graph shows the energy distribution curves of reactant molecules at two different temperatures and can be used to explain how an increase in temperature can affect the rate of a reaction.*

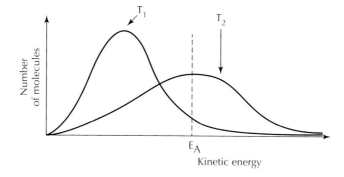

Explain why curve T₂ shows the energy distribution of the reactant molecules at a higher temperature.

The answer would require you to link the curve T_2 to higher temperature. A model answer might be:

At high temperatures, more particles have energies greater then the activation energy. Curve 2 shows more particles with energy greater than the activation energy, therefore shows the reactant molecules at the higher temperature.

Q The reaction that produces the solid sodium hydrogencarbonate involves the following equilibrium:

$$HCO_3^-(aq) + Na^+(aq) \rightarrow NaHCO_3(s)$$

Brine is a concentrated sodium chloride solution.

Explain fully why using a concentrated sodium chloride solution encourages production of sodium hydrogencarbonate as a solid.

A In this question you are asked to '**Explain fully**'. It isn't sufficient to state what happens, you need to state why the change happens.

Explanation
You are told in the question that adding brine encourages the formation of sodium hydrogencarbonate. This means that adding sodium hydrogencarbonate shifts the equilibrium to the right.

The first point that needs to be made is that adding brine increases the concentration of sodium ions, $Na^+(aq)$. Increasing the concentration of sodium ions increases the rate of the forward reaction, the reaction producing sodium hydrogencarbonate. The rate of the forward reaction is now greater than the rate of the back reaction, meaning that the equilibrium will move to the right, encouraging the production of sodium hydrogencarbonate.

Top tip

When answering questions involving equilibrium situations, always consider how any change will affect the rate of the forward and back reactions. You can then state how the equilibrium position will shift.

Remember...

Concentration
Increasing the concentration of a species in an equilibrium mixture will increase the rate of the reaction involving that species as a reactant. This is because increasing the concentration will result in more collisions and more successful collisions involving the species.

Temperature
Increasing temperature favours the endothermic reaction, the reaction using up energy. Decreasing temperature favours the energy-producing, exothermic reaction.

Pressure
Increasing pressure favours the reaction that results in fewer moles of gas. Decreasing the pressure favours the reaction that will result in more moles of gas.

Equilibrium is an area that some candidates find difficult. Although this next example doesn't have a command word, it can be used to illustrate some of the points we need to remember about equilibrium.

Q Tetrafluoroethene, C_2F_4, is produced in industry by a series of reactions.

The final reaction in its manufacture is shown below.

$$2CHCIF_2(g) \rightarrow C_2F_4(g) + 2HCI(g)$$

The graph shows the variation in concentration of C_2F_4 formed as temperature is increased.

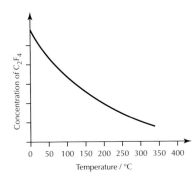

What conclusion can be drawn about the enthalpy change for the formation of tetrafluoroethene?

A In this question you have to analyse the information you are given and then come to a conclusion.

As the temperature increases, the concentration of the tetrafluoroethane decreases. This means that increasing the temperature increases the rate of the back reaction more than the rate of the forward reaction, causing the equilibrium to move to the left and decreasing the concentration of tetrafluoroethane.

Increasing the temperature favours the reaction that uses up energy – it is endothermic. The back reaction is therefore endothermic meaning that the forward reaction, the reaction for the formation of tetrafluoroethane, is exothermic. Formation of tetrafluoroethane would be encouraged by carrying out the process at low rather than high temperature.

The equilibrium in the question can also be used to illustrate the points relating to pressure and concentration we need to remember.

Looking at the equilibrium equation we can see that there are more moles of gas on the right hand side than the left (3 moles on the right and 2 moles on the left). The reaction would therefore be carried out at low pressure.

Another way to encourage the production of tetrafluoroethane would be to remove $HCl(g)$. This again would slow the back reaction and push the equilibrium position to the right-hand side.

compare – you must demonstrate knowledge and understanding of the similarities and / or differences between things.

> *Example:*

geranyl acetate   limonene

> *By comparing the structures of geranyl acetate and limonene, suggest why geranyl acetate is likely to be less volatile than limonene.*

The answer needs to relate to the structures. Volatility relates to strength of intermolecular forces, which in turn relate to polarities of bonds in molecules.

To gain the mark you need to spot that geranyl acetate contains a carbon-to-oxygen double bond making the molecules more polar than those of limonene and hence will contain stronger intermolecular forces and therefore is likely to be less volatile.

complete – you must finish a chemical equation or fill in a table with information.

> ***Example:*** *Potassium permanganate reacts with hydrogen peroxide. The equation for the reaction which takes place is:*

$$5H_2O_2(aq) + 2MnO_4^-(aq) + 6H^+(aq) \rightarrow 5O_2(g) + 2Mn^{2+}(aq) + 8H_2O(\ell)$$

> *Complete the ion-electron half equation for the conversion of $MnO_4^-(aq)$ to $Mn^{2+}(aq)$.*

$$MnO_4^-(aq) \rightarrow Mn^{2+}(aq)$$

A correctly completed equation would be required to gain the mark.

$$MnO_4^-(aq) + 8H^+(aq) + 5e^- \rightarrow Mn^{2+}(aq) + 4H_2O(\ell)$$

determine or **calculate** – you must determine a number from given facts, figures or information.

> ***Example:*** *Ethene, $C_2H_4(g)$, can be converted into ethane-1,2-diol, $C_2H_6O_2(g)$. The overall equation for the process is shown.*

$C_2H_4(g)$	+	$H_2O(g)$	+	$\frac{1}{2}O_2(g)$	$\rightarrow$	$C_2H_6O_2(g)$
mass of 1 mole						mass of 1 mole
= 28 g						= 62 g

> *112 tonnes of ethene produces 216 tonnes of ethane-1,2-diol.*

> (i) *Calculate the theoretical yield of ethane-1,2-diol.*

> (ii) *Calculate the percentage yield of ethane-1,2-diol.*

Note: In a key area assessment, all that would be required would be the correctly calculated answers to each part: (i) 248 tonnes; (ii) 87·1%.

In a prelim or the final exam the question might simply be, '*Calculate the percentage yield of ethane-1,2-diol*' for 2 marks.

It would be important therefore to show your working clearly, since partial marks could be awarded in the event of you failing to correctly calculate the answer if your method was correct.

draw – you must draw a diagram or structural formula.

> **Example:** *A dipeptide has the following structural formula.*

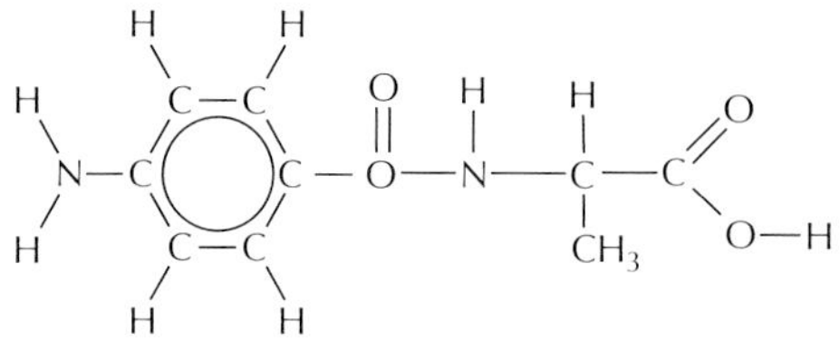

Hydrolysis of the dipeptide produces two amino acid molecules. One of these molecules has the following structure.

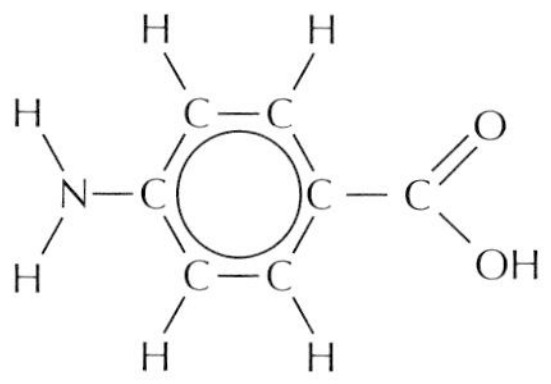

Draw a structural formula for the second amino acid molecule.

When the peptide link is broken, the two fragments formed are:

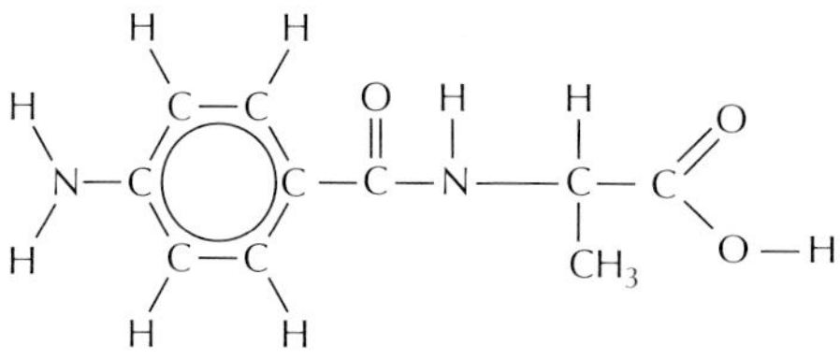

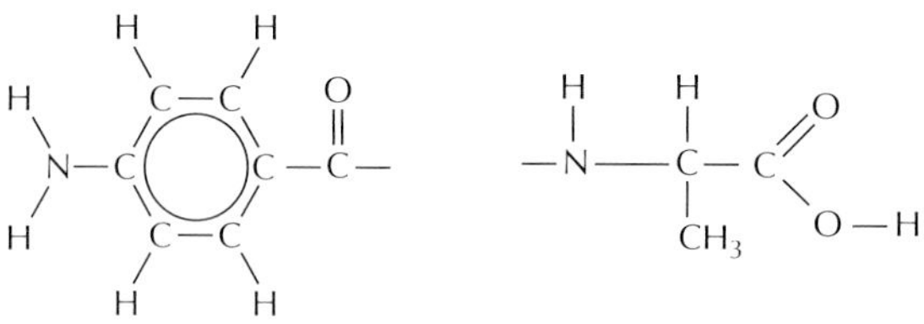

Your answer should therefore show the structural formula below, which is formed by adding a hydrogen to complete the amino group.

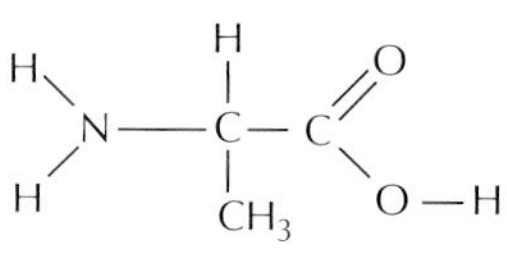

suggest – you must apply your knowledge and understanding of one area of chemistry to a new situation.

> **Example:** *Carbonyl compounds can be identified by reacting the carbonyl compound with 2,4-dinitrophenylhydrazine.*

> $R = $ *hydrogen or an alkyl group such as* CH_3, C_2H_5, *etc*

> *Suggest a name for the type of reaction that takes place.*

Although 2,4 dinitrophenylhydrazine will not have been mentioned in the course, examination of the equation shows that the carbonyl compound and the 2,4-dinitrophenylhydrazine join by eliminating a small molecule (water). This is an example of a 'condensation reaction'.

In addition to the above command words, the command words **predict** and **evaluate** are used when assessing problem-solving skills.

predict – you must suggest a property of a substance or what may happen based on available information.

> **Example** *The table shows the enthalpies of combustion of methanol, ethanol and propan-1-ol.*

Alcohol	Enthalpy of Combustion / kJ mol^{-1}
methanol	−726
ethanol	−1367
propan-1-ol	−2021

> *Predict the enthalpy of combustion of butan-1-ol.*

Butan-1-ol is the next member of the straight chain primary alcohols. Looking at the difference in enthalpy of combustion between methanol and ethanol (641) and ethanol and propan-1-ol (654), it is reasonable to predict a similar difference between propan-1-ol and butan-1-ol. Or you may think the difference might be slightly greater (641, 654... perhaps 667?). Any answer in a range from -2660 to -2690 would likely be accepted for the mark.

The value given in literature is -2670 kJ mol^{-1}, that is, 649 kJ mol^{-1} greater than that of propan-1-ol.

evaluate – you must make a judgement based on criteria such as suggesting why an experiment is carried out in a particular way or how an experiment might be improved.

> **Example** *The stock solution was prepared by adding 1·00 g of zinc metal granules to 20 cm³ of 2 mol l⁻¹ sulfuric acid in a 1000 cm³ standard flask.*

$$Zn(s) + H_2SO_4(aq) \rightarrow ZnSO_4(aq) + H_2(g)$$

> *The flask was left for 24 hours, without a stopper. The solution was then diluted to 1000 cm³ with water.*

> *Explain fully why the flask was left for 24 hours, without a stopper.*

In the above question candidates were asked to consider why a reaction was carried out in a particular way.

The answers that were required were:

- To ensure all the zinc had reacted (as the reaction proceeds and the acid is used up the reaction becomes very slow).
- To allow the hydrogen gas that is given off during the reaction to escape. (The equation gives the clue here.)

So what should you know?

A brief outline of what you should know and be able to do when asked to make accurate statements in each key area is given below. You can use this as a checklist guide when preparing for key area assessments or when revising for prelims and the final SQA exam.

Chemical Changes and Structure

The three key areas associated with Chemical Changes and Structure are:

- Controlling the Rate
- Periodicity
- Structure and Bonding

Key Area – Controlling the rate

Questions in this key area would relate to knowledge and understanding of:

- *Collision theory, explaining changing rates of reaction and activation energy.*
- *Relative rate of reaction.*

When particles collide, reactions can take place if the particles collide with an appropriate orientation and with sufficient energy to overcome repulsive forces.

The **energy required for reaction to take place** is termed the **activation energy.**

The rate of reaction depends on the number of successful collisions.

Relative rate: Since rate is a change in a quantity over time, rate is inversely proportional to time, that is, Rate α 1/t. 1/t can therefore be used as a measure of rate and is termed 'relative rate'.

Top tip

Be able to use relative rate from a graph to work out the time taken for a reaction.

> **Example** *From the graph the relative rate for the reaction at 55 oC is 0·072 s–1. This allows the time for reaction at 55 °C to be calculated.*

$$Zn(s) + H_2SO_4(aq) \rightarrow ZnSO_4(aq) + H_2(g)$$

$$\text{Relative rate } (1/t) = 0\cdot072 \text{ s}^{-1}$$

$$t \text{ (time for reaction)} = \frac{1}{0\cdot072} \text{ s}$$

$$= 13\cdot9 \text{ s}$$

Reaction profiles – potential energy diagrams, energy pathways, activated complex, activation energy and enthalpy changes.

Being aware that the course of a reaction in terms of energy change can be represented graphically.

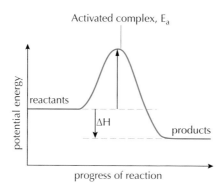

Example

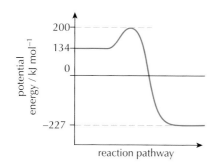

$$Ea = 200 - 134 = 66 \text{ kJ mol}^{-1}$$
$$\Delta H = -227 - 134 = -361 \text{ kJ mol}^{-1}$$

- ***Catalysts – reaction pathway, activation energy.***

Understanding how catalysts work and the effect that catalysts have for activation energy in a reaction.

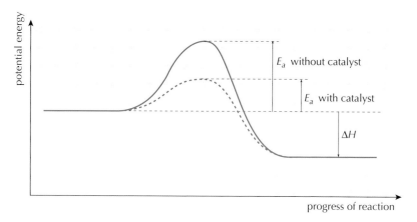

Potential energy diagram comparing Ea for a catalysed and uncatalysed reaction

- *Energy distribution diagrams showing effect of temperature changes on successful collisions. The effect of temperature on the reaction rate in terms of kinetic energy of particles.*

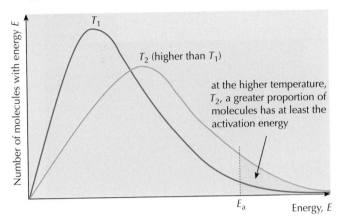

Energy distribution diagram for a reaction at temperatures T_1 and T_2

The average kinetic energy of particles in a system is proportional to temperature. The kinetic energy of particles can range from zero to very high. Increasing the temperature increases the average kinetic energy of particles meaning that there will be more with very high kinetic energy and therefore increasing the chances of collisions taking place with energy greater than the activation energy.

Key area – Periodicity
Periodic table – the first 20 elements in the periodic table are categorised according to their bonding and structure. Periodic trends and underlying patterns and principles.

Understanding of the principles used to construct the periodic table, that is, that elements are positioned in the periodic table according to their atomic structures and properties.

Detailed knowledge of the first 20 elements is required.

The first 20 elements in the periodic table can be categorised according to bonding and structure:

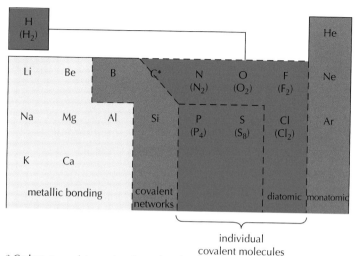

Bonding and structure within the first 20 elements

* Carbon can exist as networks and molecules

Covalent radius, ionisation energy, electronegativity and trends in groups and periods, related to atomic structure.

Periodicity describes the regular pattern of properties of the elements. Understand that trends in properties relate to nuclear attraction for outer-shell electrons and this relates to nuclear charge atom size and screening effects of inner-shell electrons.

The first ionisation energies for some elements in periods 2 to 5:

Li	Be	B	C	N	O	F	Ne
526	905	807	1090	1410	1320	1690	2090
Na	Mg	Al	Si	P	S	Cl	Ar
502	744	584	792	1020	1010	1260	1530
K	Ca	Ga	Ge	As	Se	Br	Kr
425	596	577	762	953	941	1150	1350
Rb	Sr	In	Sn	Sb	Te	I	Xe
409	556	556	715	816	870	1020	1170

—————— overall increase along period ——————→

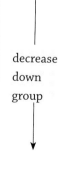

decrease down group

Key area – Structure and bonding

Bonding continuum.

Understanding that bonding is on a continuum and that bonding type changes gradually from covalent, where atoms have the same electronegativity, to ionic, where there is a large difference in electronegativity.

Polar covalent bonds and their position on the bonding continuum, dipole formation and notation $\delta+ \, \delta-$, e.g. $H^{\delta+} Cl^{\delta-}$

Polar covalent bonds form when bonding atoms have different attractions for shared electrons. Molecules with polar covalent bonds can have permanent dipoles leading to polar molecules.

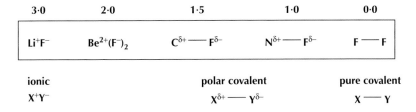

The bonding continuum – the numbers are the difference in electronegativity between the bonded atoms.

Intermolecular forces, called vdW forces, include London dispersion forces and permanent dipole-permanent dipole interactions. Hydrogen bonding and the resulting physical properties including solubility.

The term 'van der Waals' forces' encompasses the different types of attractions that exist between atoms and molecules and include London dispersion forces (temporary-temporary dipole attractions) and permanent dipole-permanent dipole interactions. Hydrogen bonding is a very strong form of permanent dipole-permanent dipole interaction.

Understand how the different types of intermolecular force arise and the impact that these forces have on properties such as melting points, boiling points and solubility.

Br —— Br ······ Br —— Br

Br —— Br   London
            dispersion
            force

mp = −7 °c

$I^{\delta+}$ —— $Cl^{\delta-}$ ····· $I^{\delta+}$ —— $Cl^{\delta-}$

$I^{\delta+}$ —— $Cl^{\delta-}$ permanent
                dipole
                interaction

mp = 27 °c

Iodine monochloride has permanent dipole–permanent dipole interactions between molecules which are much stronger than London dispersion forces.

Water contains hydrogen bonds, the strongest type of intermolecular attraction.

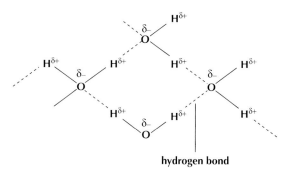

Hydrogen bonding in water.

Nature's Chemistry

There are seven key areas in Nature's Chemistry. A brief outline of what you would be expected to know for each of the key areas is given below.

Key area – Esters, fats and oils
Esters – naming, structural formulae and uses.
Recognise the ester link in compounds. Name and draw esters, given the names or formulae of the parent carboxylic acids and alcohols. Understand that ester formation is an example of a condensation reaction. Know that hydrolysis is the opposite process to condensation. Name and draw the products of hydrolysis of an ester.

ethanoic acid   methanol   methyl ethanoate   water

Fats and oils as a source of energy.
Fats and oils are naturally occurring esters. As well as being concentrated sources of energy, fats and oils are essential for the transport and storage of fat-soluble vitamins in the body.

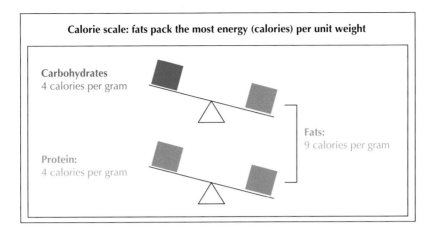

Fats and oils, esters' condensation and hydrolysis reactions.

Hydrolysis of fats and oils gives propane-1,2,3-triol and mixtures of fatty acids.

propane-1.2.3-triol
(glycerol)

the fatty acid stearic acid (octadecanoic acid)

Saturated and unsaturated fats and oils.

The fatty acids chains in fats and oils can be saturated, unsaturated or polyunsaturated.

The composition of fats and oils:

Fat or oil	Saturated fatty acids		Unsaturated (one C=C double bond)		Poly-unsaturated
	C_{16}	C_{18}	C_{16}	C_{18}	C_{18}
lard (pork)	28–30%	12–18%	1–3%	41–48%	6–7%
butter	23–26%	10–13%	5%	30–40%	4–5%
whale	11–18%	2–4%	13–18%	33–38%	
olive	5–15%	1–4%	0–1%	69–84%	4–12%
coconut	4–10%	1–5%		2–10%	0–2%

Melting points of oils and fats, through intermolecular bonding.

The melting points of fats and oils depend on intermolecular forces that in turn are related to how fat and oil molecules are able to pack due to the degree of unsaturation of the fatty acid chains.

(See Unit 2·1 of Leckie & Leckie CfE Higher Chemistry Student Book.)

Key area – Proteins
Amino acids, dietary proteins, condensation reactions to make proteins and amide / peptide link.

Proteins are the major structural materials of animal tissue. Proteins are made from amino acid units. Proteins are recognised from the amide links between amino acid units in protein chains. The amide links in proteins and peptides are referred to as peptide links.

Amino acids joining to form proteins are examples of condensation reactions.

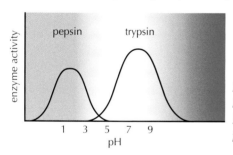

peptide link

Enzymes as biological catalysts, digestion, enzyme hydrolysis of dietary proteins.

Proteins are also involved in the maintenance and regulation of life processes. Enzymes are biological catalysts and are themselves proteins. Digestion of dietary proteins by enzymes is an example of hydrolysis and gives amino acids.

Pepsin and trypsin are enzymes produced in the body and are involved in the digestion of food. Pepsin works best at a low pH; the optimum pH for trypsin activity is at neutral pHs.

Key area – Chemistry of cooking
For aldehydes and ketones: carbonyl functional group, structural formulae and molecular formulae.

Aldehydes and ketones contain the carbonyl functional group. In aldehydes the carbonyl group is on an end carbon. In ketones the carbonyl functional group is on a carbon within the carbon chain (not on an end carbon).

Draw structural and molecular formulae and give systematic names for carbonyl compounds and their isomers with up to eight carbon atoms in their longest chain.

carbonyl groups

4-methylpentanal
(a branched aldehyde)

pentan-3-**one**
(a ketone)

Many of the flavour and aroma compounds in foods are aldehydes.

The functional groups present in flavour compounds affect the solubility, boiling points and volatility of the compounds.

Oxidation reactions of aldehydes and ketones.

Oxidation reactions involving acidified dichromate, Fehling's solution and Tollens' reagent can be used to distinguish aldehydes and ketones.

Effect of heat on proteins, denature of proteins.

Heating can denature proteins. During the process, internal hydrogen bonds holding protein chains in spirals and sheets are broken.

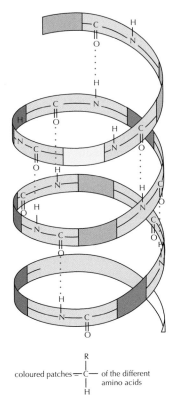

coloured patches = $\overset{\displaystyle R}{\underset{\displaystyle H}{\overset{|}{\underset{|}{C}}}}$ — of the different amino acids

A spiral chain in a protein.

133

Key area – Oxidation of food

Alcohols: for compounds with no more than eight carbon atoms in their longest chain.

Naming of branch-chained alcohols given their structural formulae and drawing structural formulae given systematic names. Naming and drawing of isomers. Know that diols contain two hydroxyl functional groups and triols three hydroxyl functional groups.

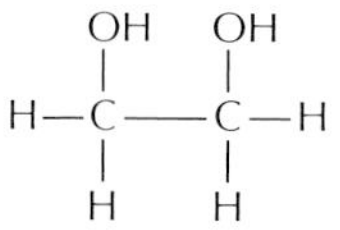

Ethane -1,2-diol

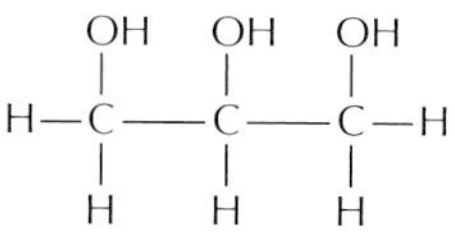

Propane -1,2,3-triol

The effect of hydrogen bonding on the properties of diols and triols.

Classification of alcohols as primary, secondary and tertiary according to the position of the hydroxyl group in the molecule.

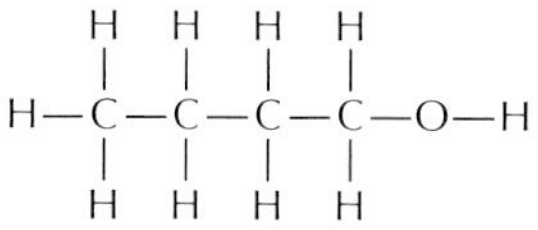

primary

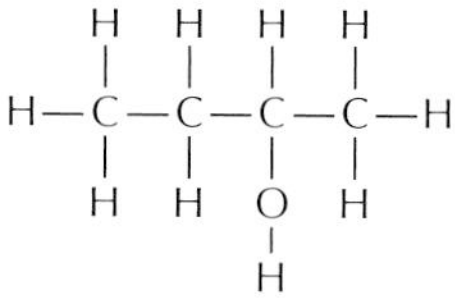

secondary

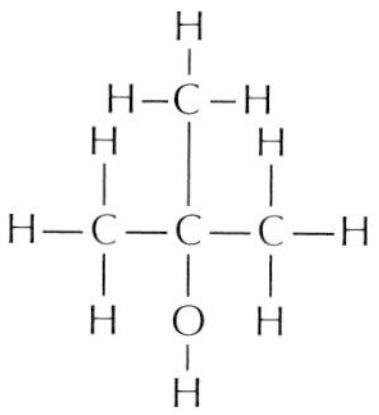

tertiary

Oxidation of primary alcohols gives aldehydes with further oxidation producing carboxylic acids. Oxidation of secondary alcohols produces ketones.

Common oxidising agents used in the lab to oxidise primary and secondary alcohols include copper(II) oxide and acidified potassium dichromate.

Remember an increase in the oxygen-to-hydrogen ratio indicates that oxidation has taken place. A decrease indicates reduction.

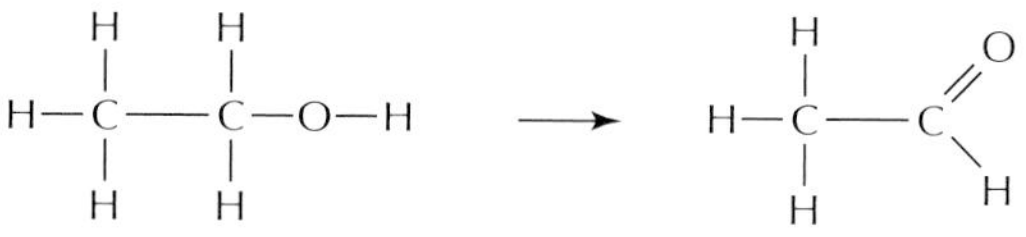

	ethanol	ethanal
oxygen : hydrogen ratio	1:6	1:4

Carboxylic acids: for compounds with no more than eight carbon atoms in their longest chain.
Naming of branch-chained carboxylic acids given their structural formulae and drawing structural formulae given systematic names. Naming and drawing of isomers. Reactions of carboxylic acids include reduction and reactions with bases to form salts.

Reaction of oxygen with edible oils.
Oxygen reacts with edible oils in food giving the food a rancid flavour.

Antioxidants. Ion-electron equations for the oxidation of antioxidants.
Antioxidants are molecules that can be added to foods and food packaging to prevent oxidation of edible oils taking place.

Ion-electron equations can be written for the oxidation of many antioxidants.

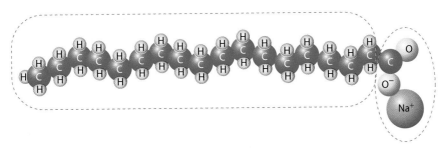

Ion-electron equation for the oxidation of vitamin C

Key area – Soaps, detergents and emulsions
Hydrolysis of esters.
Alkaline hydrolysis of fats and oils produces water-soluble ionic salts called soaps.

Structure of soap ions.

hydrophilic head

hydrophobic tail

Structure of a soap molecule

Soap dissolves in water to give soap ions, which consist of a long covalent tail that is hydrophobic (water-hating), and an ionic head that is hydrophilic (water loving).

Cleansing action of soaps.

During cleaning using soaps, the hydrophobic tails dissolve in a droplet of oil or grease, whilst the hydrophilic heads face out into the surrounding water. Agitation of the mixture results in a ball-like structure forming, with the hydrophobic tails on the inside and the negative hydrophilic head on the outside. Repulsion between these negative charges results in an emulsion being formed and the dirt released.

Production, action and use of detergents.

Detergents were developed as an alternative to soaps and are particularly useful in hard water areas. Detergents clean in a similar way to soaps.

Emulsion and emulsifiers and their formation and use in food.

An emulsion is formed from two liquids that do not normally mix when small droplets of one liquid are dispersed in the other liquid. Emulsions in food are mixtures of oil and water.

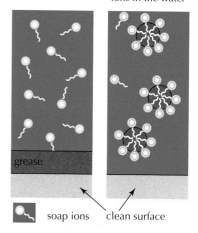

(a) soap dissolves in water forming soap ions

(b) miscelles prevented from recombining by negative charges on heads of soap ions in the water

soap ions   clean surface

Cleaning action of soap

Emulsifiers are molecules that prevent the liquids of an emulsion separating. Emulsifiers for use in food are commonly made by reacting edible oils with glycerol.

Emulsifier molecules contain hydroxyl groups that are hydrophilic and hydrocarbon portions that are hydrophobic.

$$CH_3(CH_2)_{14}C(O)O-CH_2$$
$$CH_3(CH_2)_{14}C(O)O-CH$$
$$CH_2-O-P$$

Lecithin, an emulsifier commonly used in food production.

Key area – Fragrances

Essential oils from plants, properties, uses and products.

Essential oils are concentrated extracts of the volatile, non-water soluble aroma compounds from plants. Essential oils are widely used in perfumes, cosmetic products, cleaning products and as flavourings in foods. Essential oils are mixtures of organic compounds.

Terpenes functional group, structure and use. Oxidation of terpenes within plants.
Terpenes are key components in most essential oils. Terpenes are unsaturated compounds consisting of joined isoprene (2-methylbuta-1,3-diene) units. They are components in a wide variety of fruit and floral flavours and aromas.

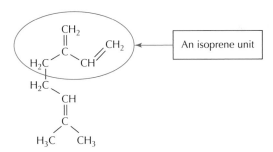

Structure of isoprene (2-methylbuta-1,3-diene)

Hint: When trying to identify an isoprene unit in a terpene look for five carbons joined in a Y shape.

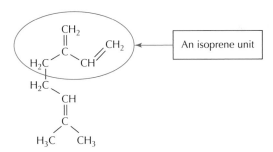

Terpenes can be oxidised within plants to produce some of the compounds responsible for the distinctive aroma of spices.

Key area – Skin care
The damaging effect of ultraviolet radiation (UV) in sunlight on skin and the action of sun-block. Formation of free radicals in UV light.
Ultraviolet radiation (UV) is a high-energy form of light present in sunlight. Exposure to UV light can result in molecules gaining sufficient energy for bonds within the molecules to be broken. Exposure to UV light can cause sunburn and also contributes to ageing of the skin. Sun-block products prevent UV light reaching the skin.

Structure, reactivity and reactions of free radicals.

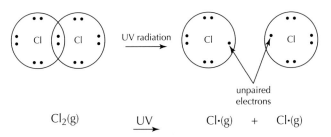

Formation of chlorine free radicals

When UV light breaks bonds, free radicals are formed. Free radicals have unpaired electrons and, as a result, are highly reactive.

Free radicals can give rise to chain reactions. Chain reactions include the following steps: initiation, propagation and termination.

Free radical scavengers in cosmetic products, food products and plastics. Reaction of free radical scavengers with free radicals to prevent chain reactions.

Vitamins are powerful free radical scavengers that are added to many cosmetic products.

Many cosmetic products contain free radical scavengers. Free radical scavengers are molecules that can react with free radicals to form stable molecules and prevent chain reactions. Free radical scavengers are also added to food products and to plastics.

Chemistry in Society

This unit contains the majority of the calculations you may face in tests and exams. However the calculations can be set in contexts that relate to other areas of the course or may even relate to industrial processes that are not covered in the course.

Key area – Getting the most from reactants
Availability, sustainability and cost of feedstock(s); opportunities for recycling; energy requirements; marketability of by-products; product yield.
Industrial processes are designed to maximise profit and minimise the impact on the environment.

Factors influencing process design include:
- availability, sustainability and cost of feedstock(s)
- opportunities for recycling
- energy requirements
- marketability of by-products
- product yield.

Environmental considerations include:
- minimising waste
- avoiding the use or production of toxic substances
- designing products which will biodegrade if appropriate.

(a)

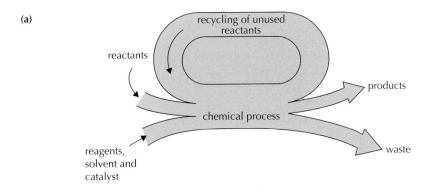

(b)

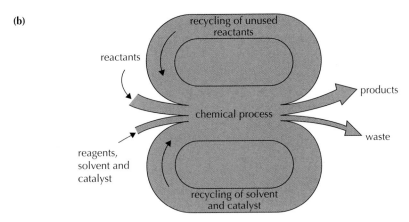

Green chemistry processes try to minimise waste through recycling of unused materials.

Balanced equations. Mole ratio(s) of reactants and product.
Balanced equations show the mole ratio(s) of reactants and products.

Determination of quantities of reactants and / or products using balanced equation, the gram formula masses (GFM), mass and moles.
Use balanced equation and gram formula masses (GFM) to perform mass-to-mass calculations.

Calculations of mass or volume (for gases) of products, assuming complete conversion of reactants.
Molar volume (in units of litres mol⁻¹) is the same for all gases at the same temperature and pressure. Use the number of moles of a gas to calculate the volume of the gas and vice versa. Use the number of moles of reactants and products to calculate the volumes of reactant and product gases.

Concentrations and volumes of solutions and / or masses of solutes.

The concentration of a solution can be expressed in mol l^{-1}. Use balanced equations in conjunction with concentrations and volumes of solutions and / or masses of solutes to determine quantities of reactants and / or products.

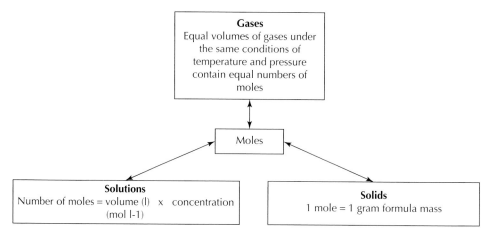

Percentage yield and atom economy.

The efficiency with which reactants are converted into the desired product is measured in terms of the percentage yield and atom economy.

Percentage yield.

Calculate percentage yields from mass of reactant(s) and product(s) using a balanced equation.

$$\text{Percentage yield} = \frac{\text{actual yield}}{\text{theortical yield}} \times 100\%$$

Given costs for feedstocks, use a percentage yield to calculate the feedstock's cost for producing a given mass of product.

Atom economy.

The atom economy measures the proportion of the total mass of all starting materials successfully converted into the desired product.

The atom economy can be calculated using the formula below in which masses of products and reactants are those appearing in the balanced equation for the reaction.

$$\% \text{ atom economy} = \frac{\text{mass of desired product(s)}}{\text{total mass of reactants}} \times 100$$

Reactions that have a high percentage yield may have a low atom economy value if large quantities of unwanted by-products are formed.

Limiting reactants and excesses.
In order to ensure that costly reactant(s) are converted into product, an excess of less expensive reactant(s) can be used.

By considering a balanced equation, identify the limiting reactant and the reactant(s) in excess.

The use of excess reactants may help to increase percentage yields, but this will be at the expense of the atom economy so a balance, that takes into account economic and environmental considerations, must be struck.

Key area – Equilibria
Reversible reactions.
Many reactions are reversible, so products may be in equilibrium with reactants.

$$\rightleftharpoons$$

The sign used to show that a reaction is reversible

In industrial processes, this may result in costly reactants failing to be completely converted into products.

Dynamic equilibrium.
In a closed system, reversible reactions attain a state of dynamic equilibrium when the rates of forward and reverse reactions are equal. At equilibrium, the concentrations of reactants and products remain constant, but are rarely equal.

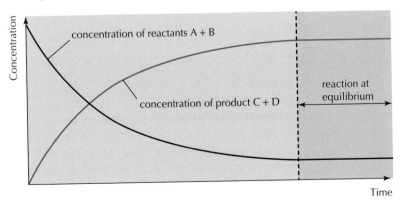

*Graph showing change in **concentration** over time for the forward and backward reactions in a reversible reaction which reaches equilibrium.*

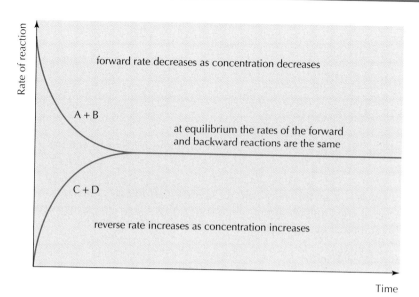

*Graph showing change in **rate** over time for the forward and backward reactions in a reversible reaction which reaches equilibrium.*

Altering equilibrium position, effect of catalyst on equilibrium and the most favourable reaction conditions.

To maximise profits, chemists employ strategies to move the position of equilibrium in favour of products. Changes in concentration, pressure and temperature can alter the position of equilibrium. A catalyst increases the rate of attainment of equilibrium but does not affect the position of equilibrium.

Predict the effects of altering pressure, altering temperature, the addition or removal of reactants or products on the equilibrium position for a given reaction.

Key area – Chemical energies
Enthalpy and industrial processes.

For industrial processes, it is essential that chemists can predict the quantity of heat energy taken in or given out. If reactions are endothermic, costs will be incurred in supplying heat energy in order to maintain the reaction rate. If reactions are exothermic, the heat produced may need to be removed to prevent the temperature rising.

Enthalpy calculations.

Chemical energy is also known as enthalpy. The change in chemical energy associated with chemical reactions can be measured. Calculate the enthalpy change for a reaction using the equation $E = cm\Delta T$ in which c = the specific heat capacity, m = mass affected by temperature change, and ΔT = change in temperature.

Enthalpies of combustion.

The enthalpy of combustion of a substance is the enthalpy change when one mole of the substance burns completely in oxygen. Enthalpies of combustion can be directly measured using a calorimeter.

Hess's law, calculation of enthalpy changes by application of Hess's law.

Hess's law states that the enthalpy change for a chemical reaction is independent of the route taken.

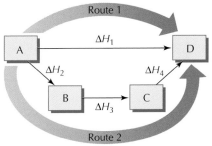

An energy cycle illustrating Hess's law:
$$\Delta H_1 = \Delta H_2 + \Delta H_3 + \Delta H_4$$

Calculate enthalpy value for reactions using Hess's law and enthalpies of combustion for common compounds that are given in data books and online databases.

Bond enthalpies – the molar bond enthalpy and mean molar bond enthalpies for molecules.

For a diatomic molecule, XY, the molar bond enthalpy is the energy required to break one mole of XY bonds.

Mean molar bond enthalpies are average values which are quoted for bonds which occur in different molecular environments.

Enthalpy changes for gas phase reactions can be calculated using bond enthalpies.

Estimate the enthalpy change occurring for a gas phase reaction using bond enthalpies to calculate the energy required to **break** bonds in reactant molecules and the energy released when new bonds are formed in product molecules.

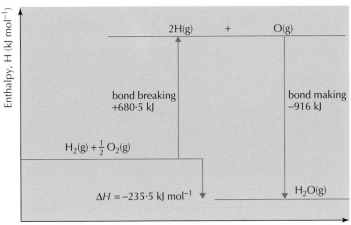

Diagrammatic representation of bond breaking and bond making when hydrogen burns

Key area – Oxidising or reducing agents

Elements, molecules, group ions and compounds as oxidising and reducing agents.
An oxidising agent is a substance which accepts electrons – an electron acceptor.

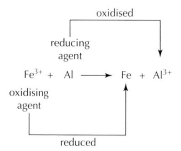

A reducing agent is a substance which donates electrons – an electron donor.

The elements with low electronegativities (metals) tend to form ions by losing electrons (oxidation) and so can act as reducing agents. The elements with high electronegativities (non-metals) tend to form ions by gaining electrons (reduction) and so can act as oxidising agents. The strongest reducing agents are found in Group 1 whilst the strongest oxidising agents come from Group 7. Compounds can also act as oxidising or reducing agents. Hydrogen peroxide is an example of a molecule which is a strong oxidising agent. Carbon monoxide is an example of a gas that can be used as a reducing agent.

Electrochemical series as reduction reactions.
The ion-electron equations in the electrochemical series are written as reductions. The electrochemical series indicates the effectiveness of oxidising and reducing agents. Strong oxidising agents are found at the bottom-left of the table and strong reducing agents are found at the top-right of the table. The dichromate and permanganate ions are strong oxidising agents in acidic solutions. Use an electrochemical series to select suitable oxidising and reducing agents to carry out oxidations or reductions.

The electrochemical series

Reaction	Strong reducing agents
$Li^+(aq) + e^-$	$\rightleftharpoons$ $Li(s)$
$Cs^+(aq) + e^-$	$\rightleftharpoons$ $Cs(s)$
$K^+(aq) + e^-$	$\rightleftharpoons$ $K(s)$
$Ca^{2+}(aq) + 2e^-$	$\rightleftharpoons$ $Ca(s)$

$Cr_2O_7^{2-}(aq) + 14H^+(aq) + 6e^-$	$\rightleftharpoons$ $2Cr^{3+}(aq) + 7H_2O(\ell)$
$Cl_2(g) + 2e^-$	$\rightleftharpoons$ $2Cl^-(aq)$
$MnO_4^-(aq) + 8H^+(aq) + 5e^-$	$\rightleftharpoons$ $Mn^{2+}(aq) + 4H_2O(\ell)$
$F_2(g) + 2e^-$	$\rightleftharpoons$ $2F^-(aq)$
Strong oxidising agents	

Uses of oxidising agents.
Oxidising agents are widely employed because of the effectiveness with which they can kill fungi and bacteria, and can inactivate viruses. The oxidation process is also an effective means of breaking down coloured compounds making oxidising agents ideal for use as bleach for clothes and hair.

Ion-electron equations for redox, oxidation and reduction processes.
Oxidation and reduction reactions can be represented by ion-electron equations. Oxidising and reducing agents appear in the equation showing a redox reaction. Balance ion-electron equations by adding appropriate numbers of water molecules, hydrogen ions and electrons. Combine ion-electron equations to produce redox equations.

Key area – Chemical analysis
Uses of chromatography. Differences in the polarity and / or size of molecules.
In chromatography, differences in the polarity and / or size of molecules are exploited to separate the components present within a mixture. Depending on the type of chromatography in use, the identity of a component can be indicated either by the distance it has travelled or by the time it has taken to travel through the apparatus (retention time).

The results of a chromatography experiment can sometimes be presented graphically showing an indication of the quantity of substance present on the y-axis and retention time on the x-axis.

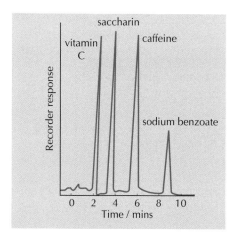

A gas-liquid chromotogram of additives in a soft drink.

Volumetric titrations, volumetric analysis for quantitative reactions.

Volumetric analysis involves using a solution of accurately known concentration in a quantitative reaction to determine the concentration of another substance. A solution of accurately known concentration is known as a standard solution. The volume of reactant solution required to complete the reaction is determined by titration. The end point is the point at which the reaction is just complete. An indicator is a substance which changes colour at the end point. Redox titrations are based on redox reactions. Substances such as potassium permanganate which can act as their own indicators are very useful reactants in redox titrations.

Calculate the concentration of a substance from experimental results by using a balanced equation.

In volumetric calculations for neutralisation and redox reactions the following relationship is used to calculate the unknown quantity:

$$\frac{\left(\text{volume} \times \text{concentration}\right)_{\text{reactant A}}}{\text{Balancing no.}_{\text{reactant A}}} = \frac{\left(\text{volume} \times \text{concentration}\right)_{\text{reactant B}}}{\text{Balancing no.}_{\text{reactant B}}}$$

Your assignment

This chapter covers:

- Introduction
- Structure of the activity
- Marking criteria
- Getting organized
- Reading for research
- Planning your experiment
- Recording results
- Referencing
- Planning your report

Introduction

The course assessment in Higher Chemistry has two components:

- the question paper
- the assignment (marked out of 20)

Your performance in the assignment is externally assessed by the SQA.

There are three phases in your assignment:

- a research phase
- an experimental phase
- a communication phase

Your assignment will normally be linked to the work you cover in the Researching Chemistry unit of the course.

The general aim of the Researching Chemistry unit is to allow you to develop key skills of scientific investigation; these are skills associated with researching scientific information and experimental skills.

You will use these skills to research the underlying chemistry of a topical issue in chemistry relating to the Higher Chemistry course and then investigate an aspect of the topical issue through planning and carrying out an experiment relating to the issue.

The work that you will carry out in the Researching Chemistry unit is likely to provide the basis for your assignment report that will be submitted to SQA.

It is really important at the outset to appreciate that although you can collaborate with others in developing research and experimental skills in the Researching Chemistry unit, and even in carrying out literature research and experimental work associated with your assignment, **you are required to write up the report for your assignment on your own** and that this will be done under closely supervised conditions.

Structure of the activity

Schools may choose to introduce the assignment in different ways. Some may choose to introduce a topic using some video clips and then have a class discussion relating to the topic. This can then give rise to focus questions which could be researched and investigated.

Researching and carrying out a practical investigation will allow you to develop the skills covered in Researching Chemistry as well as providing the basis for your assignment.

Assignment Structure

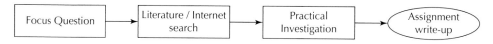

Marking criteria

It is important that you understand how your assignment will be marked by the SQA. Understanding this at the outset will help ensure you have all the material necessary when it comes to writing up your report.

Marking criteria for Higher Chemistry assignment

Criteria	What you are required to do?	Maximum mark
Aim(s)	Clearly state what it is you will investigate	1
Applying knowledge and understanding chemistry	Describe the background chemistry relating to the topic. This needs to be done at a level appropriate to Higher. **It needs to be in your own words.** Chemistry simply copied from a source will score zero.	4
Selecting information	Select information relevant to the topic from **a minimum of two sources,** one of which must be a practical activity in which you have taken an active part.	2
Risk assessment	When you carry out the practical investigation of the topic, describe any safety precautions specific to the investigation that needed to be taken.	1
Processing and presenting data / information *	You must show that you have processed raw data (this needs to be included) and present the data. **You must do both to gain these marks.**	4
Analysing data / information	You need to analyse data from at least two sources. Analysing data can include making comparisons, describing patterns and trends or discussing results.	2
Conclusion(s)	State a conclusion that clearly relates to the stated aim of the assignment.	1
Evaluation **	You must evaluate the sources you quote in your report. You need to make three statements relating to the sources. This should cover issues such as the **robustness, reliability** and **validity** of the information.	3
Presentation	Your report needs to have an appropriate title and structure. This is worth 1 mark. You must also give references for at least two sources used in the report. These must allow the assessor to be able to retrieve and check this information.	2

*Processing and presenting

Processing can include performing calculations such as calculating a concentration from a titration result or summarising information from a source that you have referenced. When summarising information you **MUST** include the raw information. This could be a printout of information from a website or a photocopy of pages of a textbook. You need to make clear within your report that the information is from a source by including a reference within the report as well as at the end of the report.

To be awarded the marks for presenting, one of the formats used to present information must be **a graph**, **a table** or **a chart**. These should have appropriate labels, headings and units. A clearly set out example of a calculation would also be a suitable presentation format. Note that a clearly set out calculation on its own would not be sufficient to gain the marks for presentation as it is not a graph, table or chart.

**Evaluation

When evaluating your investigation, marks are awarded for an appreciation of:

- validity of sources quoted – explanation of why the source might be considered to be biased / unbiased
- reliability of data / information included – data obtained from a scientific journal, etc.
- robustness of findings – information is supported by other reputable sources
- reliability and accuracy / reproducibility of experimental procedures used (experimental procedure was repeated to give results)
- evaluation of experimental procedures; factors that might be considered when evaluating experimental procedures are:
  - accuracy
  - adequacy of repetition
  - adequacy of range of variables
  - control of variables
  - limitations of the equipment
  - reliability of methods
  - sources of errors.

Getting organised

Top tip

Keep a day book.

When you come to write up your report you need to have a clear idea of the information you will include in the report. The best way to achieve this is to keep a log of all the information that you gather along the way.

A day book is simply a record of all the activities that you have been involved in. **Record everything!** This includes the websites you visit; the books and articles you have taken information from; experimental procedures; results of experiments; safety precautions; ideas about how experimental procedures could be improved, etc. Don't just write things down in any old fashion; try to achieve a structure that will allow you to retrieve information easily when creating your report.

For example, have separate sections for background chemistry, experimental procedures (this can include safety precautions and an evaluation of the procedure), experimental results, etc. You will find it much easier when creating your report if these are separate in your day book rather than jumbled together.

Remember! You are responsible for keeping your own notes. It is not the responsibility of a partner you may be working with.

Reading for research

It is likely that you will gather some of the information you include in your report from a website. There is nothing worse than going onto the internet and looking up several websites relating to a topic and then trying to remember which one had that vital piece of information you now want to include.

For all the sites you visit, note the web address and make a summary of the relevant information that you obtain there. If it contains a lot of information, bookmark the site.

Remember! You can't copy information directly from a source. That doesn't show understanding. **You need to be able to summarise information.** This can be very challenging and is a skill that **you should practise before writing up your report.**

Planning your experiment

One of the data sources you use for your report must be from practical work in which you have taken an active part.

You are not required to give a detailed description of the procedure within your report but you are required to note safety precautions and may include an evaluation of the experimental procedure in the overall evaluation of the assignment. For this reason you should always attach any experimental procedure used as an appendix to your report.

It is taken for granted that general safety precautions such as the wearing of safety glasses will be followed. The mark relating to safety is obtained by demonstrating knowledge of the hazards associated with chemical and experimental procedures and how risks associated with their use can be minimised.

Remember! You can discuss the details of any experimental procedure with your teacher and you may carry out a procedure with other students in your class.

Recording results

You should record results of any experiments in your day book. When you record your results, think about how you will process the results and how you will present raw results and processed results in your report.

Remember! When presenting information **one of the formats used** must be either **a graph, a table** or **a chart.**

Referencing

The references to at least two sources you use to produce your report need to be given in sufficient detail to allow them to be retrieved by a third party. It is good practice when using information within the report to note the reference in some way. However, **at least two references must be given in full at the end of the report.**

Full referencing means:

- for a website, the full URL address – typing it in should take you to the webpage from which information has been taken
- for a book, the book title with ISBN number or version / edition number, author and page number
- for an experiment, the title and aim of the experiment.

Remember! The last thing in your report should be references to at least two sources used in your report.

Planning your report

Writing up the report is the culmination of your assignment. You are not given a template to follow and **remember** you will be writing up your report under closely supervised conditions.

You need to have a clear idea of how you will structure your report and what you intend to write for each section before you come to write up the report. Clear notes in your day book will be key to achieving this.

It is expected that reports should be 800–1500 words, excluding tables, charts and diagrams.

You are not allowed to use a draft to write up your report but you can still practise the different aspects you might want to include in your report, such as practising writing a summary of the underlying chemistry related to the topic or setting out an example of calculations you use or drawing a graph you want to include.

Many schools will suggest that candidates structure their reports in a particular way. A suitable structure for your report would be:

Title	This should be appropriate to the assignment.
Aim	This can be the same as the title but **must be stated separately from the title.** It does not need to include the word 'aim', for example, 'To compare the vitamin C content of fresh and frozen vegetables' would be a suitable aim.
Description of underlying chemistry	A passage outlining chemistry relevant to the assignment. This should use terms and ideas at a depth appropriate to Higher Chemistry. This might include formulae, equations, properties of substances, etc.
Raw data information	Information from the literature sources you have chosen to use. The title and aim of the experiment you carried out and the raw data you obtained.
Processed data	Evidence that the raw data has been used either in summary form or to calculate other quantities.
Presenting processed information	Using graphs, tables and charts to convey processed information.
Analysing data / information	Identifying trends and relationships suggested by the processed data.
Conclusion	Stating a conclusion that clearly links to the aim and is supported by your research.
Evaluation	Assessing your assignment and conclusion in terms of the validity of sources, reliability of data, robustness of findings and the accuracy of experimental procedures used.
References	Full references for at least two sources.
Appendices	Any print-outs from webpages used to source information or used to summarise information; any details of experimental procedures, etc.

Following a set pattern like the one outlined above or one suggested to you by your teacher or lecturer will ensure you cover all aspects required in your report and will give you the best chance of achieving your full potential in this component of the course assessment.

Preparing for your final exam

This chapter covers:

- In preparation
- And on the day!

Hopefully this book has proved to be a useful aid to your studies. The aim has been to improve your understanding of what is required to answer questions in areas of the course that candidates find demanding. The different chapters should have complemented the work you have done in your Higher class. However, don't just put this book to one side now that you are reaching the end, go back and reread each chapter from time to time throughout your revision programme. If you follow the advice you will hopefully improve your skills and increase your chances of boosting your grade.

And now to finish, some final words of wisdom to help you be best prepared for your final exam.

In preparation

We all have different preferred techniques for studying. Some like music; some like absolute quiet; some like to sit at a desk; some are happy to lounge in a chair; some like to make summary notes; some like to just reread their notes. Whatever your preferred style there are things to be aware of.

The science says the two most important factors in gaining understanding are:

- spreading your revision over a long period of time (so don't leave it to the night or week before the exam)
- self-testing on a regular basis

So here are my top 10 bits of advice for being prepared for your final exam.

10 top tips for revision

1. First of all - plan your revision programme

Research shows that a technique known as 'distributed or spaced practice' is the most effective way of learning and deepening our understanding. This involves carrying out your revision over a number of short sessions spread over a longer period of time with regular testing rather than have long sessions in a short period of time just before the exam.

2. Use flash cards, post-its and memory maps

These are useful tools to help you self-test. Again, research shows that regular self-testing brings about the greatest improvement in performance. **This isn't about sitting down and doing lots of past papers** although this is necessary as well. You can test yourself on small areas of the course every day. You can use flash cards to learn things like definitions – write the concept on one side and the definition on the other. If you carry your flash cards in your pocket you can leaf through them if you are on a bus. Stick up post-it notes with key bits of information around the house – on the bathroom mirror, on the door of the fridge, beside the kettle. Draw memory maps and have them on a bedside table. When you see them, cover them up or cover bits of them with your hand and try to remember what is written down. Simple things like this can lead to big improvements.

3. Use highlighters

These **identify key words or sections** in your notes. A simple thing like highlighting a word or underlining it in colour can help you focus in on what is important on a page when reading through your notes. Remember they are your notes so you can do what you like with them to make things clear to you.

4. Do plenty of past papers

This includes past papers from the old Higher or Higher (Revised). There will be questions that you won't be able to do because the courses are different but many of them, particularly multiple choice, will be similar to the ones you face. You need to familiarise yourself as much as possible with the style of questions you are likely to face.

5. Don't put off getting started

If you are studying on a weekend day or a day during your holidays, **get started** your studying **in the morning**. We all find it too easy to put off getting started on tasks. It's easy to find excuses and think 'I'll have plenty of time later'.

6. Take breaks when you are studying

Again, research shows our concentration can be high for short periods of time but after a while we tend to drift away and lose concentration. For that reason we should structure our days with plenty of breaks. The breaks will also allow you to catch up on texts that might have come in and find out what's going on with your friends, etc. They might stop you being tempted to check your phone every few minutes.

You might structure a day's revision something like this:

- 09:00-09:30 Chemistry
- **Break 10 mins**
- 09:40-10:10 Chemistry
- **Break 10 mins**
- 10:20-10:50 Maths
- **Break 10mins**
- 11:00-11:30 Maths
- **Break 30 mins**
- 12:00-12:30 Geography
- **Break 10 mins**
- 12:40-13:10 Geography
- **Break 50 mins**
- 14:00 -14:30 Chemistry
- **Break 10 mins**
- 14:40 -15:10 Chemistry
- **Break 10 mins**
- 15:20-15:50 Maths
- **Break 10 mins**
- 16:00-16:30 Maths

7. Take exercise

You can feel sluggish after a day's study. Exercise can reinvigorate you. So do something physical: go to the gym; go swimming; take a jog. Exercise causes our bodies to release chemicals called endorphins. These help improve how we are feeling.

Regular exercise has been proven to:

- reduce our stress levels
- make us less anxious
- make us feel good about ourselves
- improve our sleep.

8. Get the right amount of sleep

Just as important as getting exercise is getting enough sleep. Getting into a regular sleep pattern will help you wake refreshed and able to cope with the demands of an intense study programme.

9. Reward yourself

It's important to get a study / leisure balance. There's nothing wrong with having the night off to meet up with friends to watch a movie or just catch up. If you are working hard then you can feel good about having a break from studying, as long as you don't play too hard!

10. Think positive

After everything is over you'll want to be able to look back and feel you did your best whatever grade you have achieved.

'The most painful thing to experience is not defeat but regret.'

Leo Bascaglia

Remember, not everyone will get 'A's. However, also remember that getting 'A's is no guarantee of future success. The good work ethic that you develop through being disciplined in your studies will definitely stand you in good stead in your future, whatever the results you achieve in your exams.

And on the day!

1. Come prepared

Come prepared, not only having completed your revision but with the physical tools needed to complete your paper. It may seem trivial but too often candidates turn up with just one pen and find 15 minutes into the exam that it has stopped working.

What do you need? As a minimum:

- pen and spare pen
- pencil
- ruler
- calculator (check it is working!).

2. Read the questions carefully

You need to make sure you understand what you are being asked to do.

3. Take note of the mark allocation for each question

This is particularly true for 'explain' questions, which can range from 1 to 3 marks. The number of marks is a good indicator of the level of detail you should give in your answer.

4. Make sure you work tidily and show working for your calculations

Set things out as neatly and clearly as possible. Positive marking is applied when marking scripts, that is, the intention is that markers should reward a candidate for showing understanding rather than penalise mistakes. It therefore helps you if you set things out clearly, allowing markers to see that you understand the various aspects of a question.

5. Keep an eye on your time

If you have prepared properly you should have plenty of time to complete the paper but if you feel you are under time pressure, make sure you answer the questions you know you can answer.

6. Don't feel you need to fill all the space that is provided for an answer

The amount of space given for an answer can depend on many things such as the next piece of writing that comes after the question. Not everyone works in the same way or has the same sized writing. You just have to be sure that what you are writing is appropriate to the question.

7. Check your answers and that you have answered every question

In the heat of an exam it is easy to miss a question at the top of a page or to turn over two pages instead of one. At the end, check you have completed every part of each question and reread your answers to make sure you are happy with them.

And now for the exam! If you have done the work there is nothing for you to be anxious about. You can go into the exam feeling confident that there is nothing you can be asked that you won't have the knowledge and skills to answer.

So good luck, but remember …

'Success is dependent on effort.'
Sophocles

'Recipe for success: Study while others are sleeping; work while others are loafing; prepare while others are playing; and dream while others are wishing.'
William A. Ward

'Success is achieved by those who try and keep trying.'
W. Clement Stone